Innovation and ICT in Education

The Diversity of the 21st Century Classroom

RIVER PUBLISHERS SERIES IN INNOVATION AND CHANGE IN EDUCATION - CROSS-CULTURAL PERSPECTIVE

Indexing: All books published in this series are submitted to the Web of Science Book Citation Index (BkCI), to CrossRef and to Google Scholar for evaluation and indexing.

Nowadays, educational institutions are being challenged as professional competences and expertise become progressively more complex. This is mainly because problems are more technology-bounded, unstable and ill-defined with the involvement of various integrated issues. Solving these problems requires interdisciplinary knowledge, collaboration skills, and innovative thinking, among other competences. In order to facilitate students with the competences expected in their future professions, educational institutions worldwide are implementing innovations and changes in many respects.

This book series includes a list of research projects that document innovation and change in education. The topics range from organizational change, curriculum design and innovation, and pedagogy development to the role of teaching staff in the change process, students' performance in the areas of not only academic scores, but also learning processes and skills development such as problem solving creativity, communication, and quality issues, among others. An inter- or cross-cultural perspective is studied in this book series that includes three layers. First, research contexts in these books include different countries/regions with various educational traditions, systems.. and societal backgrounds in a global context. Second, the impact of professional and institutional cultures such as language, engineering, medicine and health, and teachers' education are also taken into consideration in these research projects. The third layer incroporates individual beliefs, perceptions, identity development and skills development in the learning processes, and inter-personal interaction and communication within the cultural contexts in the first two layers.

We strongly encourage you as an expert within this field to contribute with your research and help create an international awareness of this scientific subject.

Innovation and ICT in Education

The Diversity of the 21st Century Classroom

Editor

José Gómez Galán

University of Extremadura, Spain

and

Ana G. Méndez University, Puerto Rico-USA

River Publishers

Published, sold and distributed by:
River Publishers
Alsbjergvej 10
9260 Gistrup
Denmark

www.riverpublishers.com

ISBN: 978-87-7022-198-6 (Hardback)
978-87-7022-197-9 (Ebook)

Contents

Preface

The adequate integration of information and communication technologies (ICT) in educational and training processes is one of the biggest current challenges in education. The classroom of the present differs greatly from just a few decades ago, new technological tools are completely transforming its characteristics and activities.

This internationally authored book offers a timely, effective, and practical vision of this new educational scenario. The work takes a multidisciplinary approach in looking at the problems and potential solutions that are challenged by the educational professional of the 21st century when, by necessity or obligation, they face the use of ICT in their daily tasks.

Divided into two parts, one theoretical and another practical, this book offers the highlights of the most important lines of research that are being developed today in educational technology and importantly presents the innovations which have had the most impact over recent years. From the profound transformations in the physical classroom to everything that involves new virtual scenarios, where online teaching requires innovative strategies and training processes, this book describes the diverse scenarios that ICT has generated and will continue to produce in education. It presents a new and very different type of education that can be adapted to the needs of the citizen of the digital society.

List of Contributors

Ana Isabel Gómez Vallecillo, *Santa Teresa de Jesús Catholic University of Ávila, Spain; E-mail: anai.gomez@ucavila.es*

Ana Marta Santamaría Rastrilla, *Santa Teresa de Jesús Catholic University of Ávila, Spain; E-mail: ana_rastrilla@yahoo.es*

Ángel M. Delgado-Vázquez, *Pablo de Olavide University, Spain; E-mail: adelvaz@bib.upo.es*

Antonio Luque de la Rosa, *University of Almería, Spain; E-mail: aluque@ual.es*

Aránzazu Cejudo-Cortés, *University of Huelva, Spain; E-mail: carmen.cejudo@dedu.uhu.es*

Blanca López-Catalán, *Pablo de Olavide University, Spain; E-mail: blopcat@upo.es*

Carmen María Hernández Garre, *University of Almería, Spain; E-mail: cmhgarre@ual.es*

Celia Corchuelo-Fernández, *University of Huelva, Spain; E-mail: celia.corchuelo@dedu.uhu.es*

César Bernal Bravo, *King Juan Carlos University, Spain; E-mail: cesar.bernal@urjc.es*

Cristina Lázaro Pérez, *University of Murcia, Spain; E-mail: cristina.lazaro2@um.es*

Diego Vergara, *Santa Teresa de Jesús Catholic University of Ávila, Spain; E-mail: diego.vergara@ucavila.es*

Eloy López Meneses, *Pablo de Olavide University, Spain; E-mail: elopmen@upo.es*

Emilio José Delgado-Algarra, *University of Huelva, Spain; E-mail: emilio.delgado@ddcc.uhu.es*

Esteban Vázquez-Cano, *UNED, Spain; E-mail: evazquez@edu.uned.es*

Eva Ordoñez Olmedo, *Santa Teresa de Jesús Catholic University of Ávila, Spain; E-mail: eva.ordonez@ucavila.es*

Jesús Valverde-Berrocoso, *University of Extremadura, Spain; E-mail: jevabe@unex.es*

Jose Ángel Martínez López, *University of Murcia, Spain; E-mail: jaml@um.es*

José Fernández Cerero, *University of Seville, Spain; E-mail: jfercerero@gmail.com*

José Gómez Galán, *University of Extremadura, Spain; E-mail: jgomez@unex.es & Ana G. Méndez University, Puerto Rico-USA; E-mail: jogomez@uagm.edu*

José Juan Carrión Martínez, *University of Almería, Spain; E-mail: jcarrion@ual.es*

José María Fernández Batanero, *University of Seville, Spain; E-mail: batanero@us.es*

José María Mezquita Mezquita, *Santa Teresa de Jesús Catholic University of Ávila, Spain, and IES Maestro Haedo, Spain; E-mail: unmezquita@gmail.com*

Juan Francisco Alvárez-Herrero, *University of Alicante, Spain; E-mail: juanfran.alvarez@ua.es*

María del Mar Fernández Martínez, *University of Huelva, Spain; E-mail: mar.fernandez@dstso.uhu.es*

María Naranjo Crespo, *Complutense University of Madrid, Spain; E-mail: marnaran@ucm.es*

María R. Belando-Montoro, *Complutense University of Madrid, Spain; E-mail: mbelando@edu.ucm.es*

María Rosa Fernández-Sánchez, *University of Extremadura, Spain; E-mail: mafernandezs@unex.es*

Nellie Pagán Maldonado, *Ana G. Méndez University, Puerto Rico-USA; E-mail: npaganm@uagm.edu*

Omar Ponce, *Ana G. Méndez University, Puerto Rico-USA; E-mail: um_oponce@uagm.edu*

Pilar Moreno-Crespo, *University of Seville, Spain; E-mail: pmcrespo@us.es*

Rosabel Roig-Vila, *University of Alicante, Spain; E-mail: rosabel.roig@ua.es*

List of Figures

List of Tables

List of Abbreviations

AR	Augmented Reality
CSS	Cascading Style Sheets
CERI	Centre for Educational Research and Innovation
CSCL	Computer Supported Collaborative Learning
CBT	Computer-Based Training
CREA	Community of Research on Excellence for All
CMS	Content Management System
CPD	Continuing Professional Development
DeSeCo	Definition and Selection of Competencies
LMS	E-Learning Platforms
eLearning	Electronic Learning
ETE	Emerging Technologies for Education
F2F	Face to Face Training
FAQ	Frequently Asked Questions
GSRS	Game-Based Student Response Systems
GLE	Gamified Learning Environment
GUI	Graphical User Interface
HRIS	Human Resource Information System
HTTP	Hypertext Transfer Protocol
ICT	Information and Communication Technologies
iTEC	Innovative Technologies for Engaging Classrooms
ID	Instructional Design
IAU	International Association of Universities
HEDIB	International Bibliographic Database on Higher Education
IoT	Internet of Things
L&D	Learning and development
LCMS	Learning Content Management System
LETSI	Learning Education Training Systems Interoperability
LSM	Learning Management System

LRS	Learning Record Store
LTI	Learning Tools Interoperability
LED	Light-Emitting Diode
MOOC	Massive Open Online Course
MIL	Media and Information Literacy
m-learning	Mobile Learning
Moodle	Modular Object-Oriented Dynamic Learning Environment
MCQ	Multiple-Choice Questions
NCES	National Center for Education Statistics
OCW	Open Course Ware
OER	Open Educational Resources
OSS	Open Source Software
OECD	Organization for Economic Co-operation and Development
PC	Personal Computer
PLE	Personal Learning Environments
PNG	Portable Network Graphics
PIFS	Practical Intelligence for School
PROPEL	Production, Reflection, Perception, Learning
PISA	Programme for International Student Assessment
RSS	Really Simple Syndication
R&D	Research and Development
RLO	Reusable Learning Object
SDP	School Development Program
SCORM	Shareable Content Object Reference Model
SOAP	Simple Object Access Protocol
SEN	Special Educational Needs
SAM	Successive Approximation Model
TALIS	Teaching and learning International Survey
TPeCS	Technology, Pedagogy, Content and Space
URL	Uniform Resource Locator
UNESCO	United Nations Educational, Scientific, and Cultural Organization
VLE	Virtual learnIng Environment
VR	Virtual Reality
VT	Virtual Tools
WOS	Web of Science

1

ICTs, Innovation, and Educational Research

José Gómez Galán

University of Extremadura, Spain, and Ana G. Méndez University,
Puerto Rico-USA
E-mail: jgomez@unex.es; jogomez@uagm.edu

1.1 Introduction: Educational Research as a Key to ICT Integration

The presence and importance of information technology and telematics, and in general of ICT in today's society, has led to a significant change in methodological research processes in the educational sciences in recent years. Both from the perspective of theoretical research and from field work, it has become necessary to develop new models of educational research that allow to meet the needs generated by new pedagogical and didactic environments, increasingly widespread in the international sphere and influenced by the impact of ICT. It should be borne in mind that from a traditional perspective the research developed within the educational sciences had focused on studying, analyzing, controlling and/or applying different educational phenomena that for many decades were quite homogeneous.

In most cases the research process is aimed at providing scientific interpretations based on the adequacy of educational processes to the contexts in which they were developed, i.e., social, economic, political, etc., as well as structural, within the educational systems themselves. However, the explosion of ICT during the last 30 years, that is to say, especially since the appearance of the personal computer (PC) and other personal computing systems (in addition, of course, all the dependent or developed instruments), as the main milestone of computer science, as well as the generalization and expansion a few years later at all levels, including domestic, of the network of networks (the Internet communicative paradigm) in the field of telematics (both currently complementary and main protagonists of the new digital society) have assumed that the dimensions indicated earlier, both the relationship of educational systems with other systems (social, political, economic, cultural) and intraeducational must be seriously revised from the perspective of educational research (Gómez Galán, 2012, 2017a).

The main reason, of course, is that the emergence of ICT in education has implied the existence of variables and elements unknown up until now that make the research models that we could call classics become insufficient, of course, to access an empirical development that allows the achievement of objectives that have a direct or indirect relationship with the influence or presence of these technologies in the educational world. In the context mentioned, naturally, different questions arise to which we must answer: what models are, therefore, the best ones to carry out a process of investigation in the new forms and necessities of the educational systems in which the presence of ICT is growing? How can we know the most appropriate strategies for integrating new technologies into education, according to the urgent training needs of a society that is increasingly dependent on them? How to get the implementation process to be appropriate, to be established in the correct form, manner, and times so that it can be assimilated by systems so hermetic and immovable, although apparently not presented in this way, as educational?

1.2 New Research Models in a New Reality

The answer would be, of course, to have research frameworks adapted to the new reality, that would allow the information to be effectively obtained, that the analysis and evaluation of the information be effective and accurate, and that the results allow correct and productive decisions. Conscious, therefore, that we are in a world in rapid mutation, in accelerated change, where new technologies acquire importance and power, the school, as guarantor of human evolution (together with the nature in which it is situated) and the personal development of individuals of our species, cannot remain apart from the many problems that such a transformation is producing.

It is not only a question, and in this, many authors have been focused effectively and from different points of view (Schoepp, 2005; Bingimlas, 2009; Buabeng-Andoh, 2012; Albion, Tondeur, Forkosh-Baruch & Peeraer, 2015;; Holmberg, 2017; Leahy, Holland & Ward, 2019; Williams & Beam, 2019; Kenttälä & Kankaanranta, 2020). These new and powerful tools of information and communication (the bases of all teaching process) in educational contexts, to improve, develop, or achieve didactic dynamics more suitable for the achievement of different curricular objectives (turned into educational resources of the first level). In all, what is really important in our view is that it is necessary to train the child, the adolescent, and the youth of present society, which is in integral growth and preparing for its prolonged rite of entry into the universe of adults (of decision-making and responsibility), in understanding and analyzing one of the elements that will be most important in the course of their life, communication, whose form and characteristics are gradually transformed due to the emergence of new instruments and technological means that arise, with different interests and objectives, to enrich it, to expand it, to multiply it, to diversify it, to grant it, and in the end, power (Gómez Galán, 1999, 2015a; Gómez Galán & Mateos, 2002, 2004).

Therefore it should not be a surprise to us that one of the main topics of interest that is currently generating a lively debate in the field of education is focused on research methodologies. The new panorama of ICT has not been indifferent to the models of educational research, in such a way that we are faced with the challenge of perfecting and creating new models that are more satisfying to the many questions that we have about education today. It is necessary to present new and innovative proposals that help the educational researcher to know not only what constitutes the state of the matter in this important field of educational sciences but, above all, to open a window to new possibilities and tendencies (Gómez Galán, 2016, Gómez Galán & Sirignano, 2016). We still have much to progress in this regard.

We are talking about a scientific field, such as education, which not only contributes decisively to the construction of the world of the present, but is fundamental in the future. Everything we apply today in this context will not be something that we will see immediately, but will reap its fruits in the next generation, which is currently being formed. Therefore, in education it is also necessary to know, in a multidisciplinary context, everything that affects the world today, its multiple problems, because education must be above all the engine that contributes to the improvement of society and the evolution of the human being. It must be, of course, the main tool to deal with the most important problems that we must solve. This implies, therefore, that educational research must be a global research, research not only focused on teaching- learning processes, as is often contemplated, but research that is capable of responding to the needs in the multiple fields of knowledge that have a direct or indirect relationship with our science.

We live in fluid times, to use the words of Bauman et al. (2015). These are the times in which it is necessary to look for an orientation and structures on which to settle. In spite of the great potential that the ICT offers us the really useful thing will be to maximize its benefits and to minimize its disadvantages, that of course it has them and they are very many. And in the field of educational research it cannot be otherwise. The great possibilities that ICT open for the management of huge volumes of information cannot make us lose sight of the fact that our work material is essentially the human being, society and, in general, the environment—with all forms of life which this includes—in which we are (Ponce, Pagán & Gómez Galán, 2017). There is little sense in having to carry out very complex studies, e.g., in statistics, using quantitative methodologies that would rival those employed in the field of experimental sciences, if we lose sight of the questions and needs we have today in education, which must be the ones guiding our work.

However, in the last decades and dazzled by the quality of the scientific methodologies used in the field of experimental sciences, educational sciences have begun to participate in complex methods and techniques much more focused on the data than on the essence. As we said, the use of ICT to perform statistical calculations, e.g., has meant the generalization of multiple studies of a quantitative nature that rarely contribute to the progress of what education really is, beyond the anecdote of knowing data that does little more than serve to satisfy a curiosity. At times, research models are established that are applied in multiple case studies that are clearly unnecessary because there are already enough samples to reach conclusions. But the theoretical splendor of scientific rigor—with professional connotations for researchers rather than real utility in its application—is often praised far above what the research is really giving us. Even studies that would be fully justified by qualitative methods are transformed into quantitative or, at least, mixed, the process being deductive at the service of the method rather than the objectives.

In addition, in many occasions we move through fashions, methods that have successes in a particular area but that are generalized to a whole without being accompanied by a calm reflection that justifies them. And conversely, techniques that were traditionally successful and that gave us excellent results are denounced, cornered, and forgotten. And there is an even greater problem in all these dynamics, and it is none other than the result of these investigations—irrespective of their scope—is restricted to the field of merely academic and professional, without there being a communication of results—which would imply carrying them to practice—to the educational community, to teaching professionals, to families, to students, all of them the true protagonists of what we understand as education. It has long been alerted to the problems of communication that existed in educational research (Carr, 1996); today, paradoxically, we are in a hyperconnected world, and far from being resolved they have become much stronger.

1.3 Pedagogical Methodologies and Innovation

But in our opinion there is an explanation for this, which is a doubt, and is none other than the university professionalization of pedagogical research. Far from having absolute freedom to focus their efforts on the real problems that today affect education—and as we said at the beginning are the main problems that involve our society—educational researchers are forced to investigate to publish. Only the professionals of the university world are those who are trained to know the educational methodologies, and consequently are those who work in this field of education. But their accountability is to their institutions, not to society, and they are forced to be valued and measured according to the theoretical quality of their publications, conditioned by factors of impact and rankings of powerful media companies.

Far from it, therefore the professional of the education that, in addition, is an researcher. Whoever researches in the educational field is a professional, first and foremost, of the research itself. Certainly in higher education, both dimensions would be merged. But it does not happen in other educational levels, the most important for the future of society. For research carried out in early childhood education, primary education, or secondary education are not those carried out by educational agents who work day to day and live in them, but by those university research professionals who see their methodologies conditioned by a final product, the researcharticle, which will be affected according to them by indexed impact journals (Gómez Galán, 2015b). And no matter the why, the what, and the goal. What matters is the how, the research method. For the educational article today what will be asked fundamentally will be its similarity to an article of experimental sciences, i.e., of professional researchers, and for that will require the most similar methodology possible. The problem, the question, and the concern that affects our society—what will really transform and improve our world—will be in the background. It is enough to see, globally, the themes of educational articles published in such decisive scientific impact journals, and to see if they really contribute to what education needs today. Or rather, they are conditioned by the bibliometric parameters that tyrannize and condition the freedom of the researcher, since the evaluators of the article, let's not fool ourselves, will fundamentally consider the method. Articles that are perfect methodologically, but empty of meaning and practical application.

In this context, the real professionals of education at each educational level have few possibilities to investigate. And if they do, they will be conditioned by investigative methods. If they do not use the appropriate ones—and in many cases we refer to those that we can consider in fashion—they will naturally never see their contributions published, however important they maybe. They will encounter problems in putting forward and designing their research projects. That is why training in this field is so necessary for any educational professional (Gómez Galán, 2011, 2014c). For it is advisable that whoever carries out his/her daily work at that specific level is the one who is better equipped to deal with those field studies. This is because part of a valuable information in most of the occasions is inaccessible through other research methods: your daily experience. Therefore, necessarily, this must be systematized and organized if we really want to achieve the proposed objectives.

What role does ICT play in all of this? The answer is simple: they must be at the service of the process and not be the end of it. Educational research should use these tools whenever they are necessary, but in no way condition all the research work to integrate their presence justifying it in the power they have to manage information and communication. That an investigation is perfect from a methodological perspective and that offers an overwhelming amount of data analyzed, and thanks to these new management tools does not imply that it is useful.

Therefore, ICT can greatly aid in educational research, create new methodological models, bring new social media closer to the world of education, but must always be at the service of what is really necessary to know to improve the educational processes and, ultimately, the evolution of our civilization. It is necessary to promote the training of teachers in research tasks at all levels of education, it will be necessary to facilitate research and innovation by teaching professionals and also not to put the research tasks at the service of bibliometric indicators as fundamental measures to focus on the problems that really affect education (Gómez Galán, 2015a). In this work the emergence of new technologies can contribute as an extraordinary aid to facilitate these processes, but in no way will be everything. The greatest innovations, therefore, not only have to be technical but also theoretical.

1.4 Conclusions

We are also faced with another problem, which we will elaborate on in the next section. It is produced by the splitting up in different specialties that, throughout the 20th century, occurred in the educational sciences. From the classical psychology of education, biology of education, philosophy of education, sociology of education, economics of education, history of education, theory of education, etc., up to those included in the field of didactics or in the new frameworks of educational technology, in each one of them appeared corresponding research models with preferred methodologies and extremely defined research processes, which complicated much more the theoretical bases of educational research.

Although in our case we defend a convergence of these specialties toward new holistic models, it is undoubtedly difficult to overcome what has been separating them for decades. Therefore, and as it is widespread that there are educational researchers for each of these areas, there is no doubt that it must mean that the greatest impact of ICT is precisely in educational technology, since it is its essence and *raison d'être*. It will therefore be necessary for us to focus on how the vast development of new technologies in a society has decisively influenced this specialty, educational technology and, especially, undoubtedly, in its research models.

It is from the research developed from the multiple models that can face the integration of ICT in education that innovative processes can be carried out with guarantee. It should not be forgotten that innovation is the introduction of a new element in the classroom. Sometimes it is possible to carry out an experiment and prove its effectiveness, but in other cases there must be prior guarantee that the introduction of this new element will be effective. It is a complex problem that unifies research, innovation, theory, and educational practice. All of this is united by a common nexus that is digital technologies as revolutionary elements of pedagogical frameworks.

2

Models in Educational Technology I: Molecular Models

José Gómez Galán

University of Extremadura, Spain, and Ana G. Méndez University, Puerto Rico-USA
E-mail: jgomez@unex.es; jogomez@uagm.edu

2.1 Introduction: Approach to Molecular Models

As we have presented in Chapter 1 this is the time to establish the main research models in educational technology based on the classification we have proposed. The first great dimension is the molecular models. In this case we refer to all those investigative models that focus on certain areas of educational technology, without considering the specialty from a holistic perspective and in dialogue with many other educational sciences and social sciences, especially looking for the integration of ICT in educational processes in certain methodological and didactic contexts.

The molecular models essentially contemplate these technologies and means as resources or didactic auxiliaries. We would start from the old classification that divided educational technology between "technology in education" and "technology of education" (López-Arenas, 1985; Gómez Galán & Sáenz del Castillo, 2000; Ferreiro & Di Napoli, 2006). The first one concerned the integration of various technologies into educational processes with an essentially didactic purpose, while the second focused on a theoretical concept of turning education into a technology, as understood in the field of philosophy of science. Molecular models would correspond to that first concerned dimension of the use of technologies and means in educational processes.

Indeed, in this sense, the presence of ICTs in school contexts has meant the presence of a new element that, although not radically revolutionary, as hinted at in some studies (Negroponte, 1995), should be analyzed and investigated in the new parameters in which it is located. In the case of educational research, there would be different approaches that would require particular research methodologies (curricular, psycho pedagogical, didactic, methodological, evaluative, organizational, etc.) depending on their areas of belonging to the educational sciences. As well as those that are aimed at knowing and solving the problems that arise from the direct presence of technological instruments in educational centers, both in the classroom and in the management and administration units. From this dimension, which we consider to be molecular, it would be necessary to establish the contextualization and development of the research processes that can be carried out. Naturally all systematization of these characteristics, as we will offer later, only implies an orientation to facilitate the understanding of a complex field, as in this case is educational research in ICT.

Offering classification in the field of educational research undoubtedly helps to address the selection of objectives of all research and experiences that consider the presence and development of new technologies in school spaces (Gómez Galán, 2017b). It also contributes to the centering of the lines of work, in such a way that it makes it possible to approach concepts traditionally more widespread and used in epistemology and, in particular, educational terminology. In a synthetic way, we could point out that the main function of ICT in school (although not yet fully extended), and the most complex in turn, is that of didactic resources, that is, auxiliary teachers or students. Their presence as instructional systems (e.g., for e-learning processes, m-learning, MOOC, etc.) or as an aid to administrative tasks would imply an independent study, in the first place because it is a clearly specialized field, and in very specific educational contexts (essentially in the university world or postgraduate education), or to connect, in the second, with research processes parallel to what is essentially pedagogical, and closer to management and economics.

We focus, therefore, on considering new technologies in the school, from the molecular perspective, as didactic resources, with what this implies from the educational research because we can use the methodological processes that have characterized the studies centered in the same ones since practically the origins of the education sciences. However, confronting them from this point of view presents, in turn, different readings. Basically we can say that it would be a question of considering ICT as an auxiliary of the teaching function of the teacher with the goal of encouraging the assimilation of contents by students, i.e., that facilitate and optimize the teaching-learning processes for the purpose to help in achieving different curricular objectives. However, they can also play a significant role in self- learning, motivation, the application of innovative didactic methodologies, etc.

Talking, therefore, of the new technologies as didactic resources implies alluding to their educational use, regardless of how it is, to favor or achieve learning. Due to this, the research lines that have focused on studying the appropriateness of these resources in didactic processes have tried to establish their advantages (and disadvantages) from the different paradigms of learning and knowledge with which, from fundamentally psychopedagogical perspectives, we work with today. The problem, of course, must be analyzed, since, of course, depending on how we consider human learning and the acquisition of knowledge, ICT as learning resources will acquire different functions and purposes. We must not forget, of course, that these technologies today dominate all information and communication processes in the world, and this is due to the digitization of information. All the old media and analogical supports have been replaced, in the digital paradigm, by a single medium in which all the systems of communication created by the human being from its origins converge.

2.2 The Phenomenon of Techno-Media Convergence

This phenomenon, which we have termed techno-media convergence (Gómez Galán, 2003), no longer establishes a separation between what was traditionally informatics and telematics, nor between traditional media (cinema, radio, television, newspapers, etc.) and the new ones (information and personal communication systems), they presuppose a permanent connection to information networks with the existence of a single digital medium in which all of them are merged. Tablets, smartphones, netbooks, etc., are signs of the techno-media convergence process, as we announced more than a decade ago (Gómez Galán, 1999) is not only unstoppable and that will lead us to a future of full connectivity, but they are already part of our present, and they are already the social reality in which we live.

Today ICT is not tools oriented to professional contexts: they are very powerful mass media. Taking into account this consideration, which goes unnoticed in many studies of new technologies that considered them completely independent of the rest of the media (which we understand to be incorrect), what is really characteristic of them is that they are used fundamentally today (and practically the whole scientific bibliography participates partially or completely with this vision) to favor or to obtain the learning of diverse contents (be they conceptual, procedural, or attitudinal), to help the students in the acquisition of different capacities, for, to ultimately, achieve different educational objectives. That is, they are used primarily to facilitate student learning (Gómez Galán, 2012).

As we can see, therefore, one of the main concerns that educational technologists have today is the integration of ICT as didactic resources. What is the way to do it? In this case the possibilities have been very broad but all start from the fact that it is necessary to take into account the theories of learning, on the one hand, to be the basis of work in the classroom, and the theories of communication on the other, inherent to the nature of these technologies. These theories are the basis of current molecular models that seek to achieve the stated objective.

It would take us a long time, and away from the objectives of this work, to focus on the complexity of the foundations of communication theories that have served as theoretical support for the molecular models of educational technology research, and its epistemological structure. Suffice it to say that they were developed especially from the importance that the mass media acquired in society. Classical contributions such as Lazarsfeld & Merton (1948), Shannon & Weaver (1949), Berelson (1949), Barthes (1957), Eco (1979), Mattelart (1994), Bruggeman (2008), or more recent works such as Hjorth & Hinton (2019), Ott & Mack (2020), or Isin & Ruppert (2020), which contemplate the impact of the Internet paradigm on communicative processes, provide an overview of the evolution of these theories. In the educational field, it is based on the fact that communicative social models can be adapted to the teaching-learning processes, which are, in the end, nothing more than communication processes independent of the objective pursued (in this case instructive and/or educational).

On the other hand, and in relation to learning theories, independently of those that, as we have indicated, relied on theories of communication, there were many proposals from different areas of knowledge (psychology, pedagogy, philosophy, etc.) which focused precisely on human learning processes, minimizing their dialogue with other supradimensions (social, political, economic, etc.). In this sense we can cite the two great paradigms generated during the 20th century: *behaviorism* and *constructivism*.

2.3 Main Molecular Models

The fusion of all these currents has generated the molecular models of research in educational technology, the reason why its structural bases are certainly solid. ICTs are contemplated as didactic aids with which to enhance or allow different teaching-learning processes, mainly in the classroom (but also in other formative contexts, such as distance education) that must necessarily be linked to the studies carried out for this purpose on classical media, the source of theories of communication, or the postulates of the theories of learning developed during the last century. Failure to take this into account would mean exploring unknown ways of investigation of questionable results, which is certainly dangerous given the current need for an adequate integration of new technologies in school settings. And more in a context, as referred to the techno-mediatic convergence, in which the communicative processes generated from the digital paradigm penetrate with intensity in all the parameters of learning.

Taking into account all of the above, we find ourselves at the moment of proposing the molecular methodological models that we identify, nowadays, with a greater presence in educational research in ICT. We can group them into three main blocks: evaluative studies, comparative studies, and intramedium studies and ATI research. These two last ones, which we will describe together, seem to us the paradigms of educational technology, in the molecular dimension, with more possibilities of positive results today.

We will begin with evaluative studies. In this group are all those works that individually consider the possibilities of each technology and medium, which is analyzed and evaluated from an educational perspective. We can consider the hundreds of investigations that have been carried out from this point of view during the 20th century, from the first experiences that were carried out, e.g., with the phonograph or with the cinema (already in the first decades of the last century) to the very frequent ones in the present, on the personal computer, Internet, video games, or virtual reality. All of them attempt to demonstrate (or reject) the intrinsic value of each of these means or technologies for instructional or, in general, educational tasks.

Today it is possible to affirm that these studies are only valid for the concrete contexts in which all the variables are realized, controlling and meticulous measuring all variables. Outside of these specific frames, and of course with the presence of different variables, the results cannot be transferred. Thus, while for the Internet (and the same can be said of any other means), e.g., it maybe appropriate for specific teaching-learning processes and for the achievement of delimited educational objectives, as demonstrated in a work of a very carefully planned field for correctly analyzed initial and final conditions, in other circumstances or spaces, no matter how similar they maybe (i.e., with the presence of very similar variables, but of course not quite the same) their usefulness can be clearly questioned. Therefore, we cannot say that the results offered by such research can be generalized. At most, as we have pointed out, they can determine their validity for very concrete contexts.

The lines of research that we can consider more serious and rigorous in the field of new technologies do not work with evaluative studies. Nonetheless, for a long time they were research models that enjoyed great international prestige (which even had great resonance, and which on many occasions were put as models of studies in ICT) regardless of whether they could try to generalize, in greater or smaller measure, their results or that they tried to collect many of the partial conclusions reached in molecular investigations and, seeking the coincidences in the same ones, it was tried to establish assessments on the educational quality of the technology or medium. As we will present in the next point, today, it is impossible to talk about quality in relation to its educational power (the more you can refer to its ability to transmit information, communicative perspectives, instructional characteristics, etc.). The possibilities of using a medium for educational purposes, and even for specific teaching-learning processes, will be determined by the variables present in the application context. Of course, and as we have pointed out, the analysis and evaluation of media in an independent way has not meant, nor is it supposed to be, the only line of research by specialists in educational technology and specifically media didactics (Gómez Galán, 2016) .

Comparative studies, among various media applied to some educational object, have also been extremely frequent. This was intended to establish which technology or medium is most appropriate for a given area or subject, or to achieve certain objectives. Usually, the comparison is made between a new medium and a traditional one, between new technologies, or within didactic processes in which in one case a particular medium of communication is used and in another case it is ignored, applying a methodology based on expository techniques.

However, since the first experiences at the beginning of the 20th century, it was demonstrated that what is most important is not the medium itself, but the use made of it, the characteristics of students, and other related variables (Gómez Galán, 2003). Due basically to the dialogue between social psychologists and educational media researchers, a great interest was established in developing research on social media in which their instructional power was compared (Rogers, 1986), and if this corresponded to the presence and influence that each of them has in society.

Practically from the 1920s to the 1960s these studies were very frequent, although most were characterized by leaning on very fragile theoretical concepts, developing poor experiments, and not offering significantly important results (Hartley, 1974). This also led to conflicting results, or to erroneous assumptions (Knowlton, 1964), which were collected in educational contexts, producing certain detrimental effects. Therefore, the initial question in the comparative research of technologies and media in educational settings, and which still make many specialists, i.e., which medium is the most effective for learning, would not make sense (Clark & Salomon, 1977) since it would always depend on the context. There will be much more adequate means to achieve certain objectives than others, no doubt, but will be depending on where this learning, the whole process of teaching and learning is developed.

Many authors have raised their doubts about comparative research. For example, Clark (1985) argues that in virtually all investigations what has ultimately been measured has been the influence of the media on certain academic outcomes, rather than the actual difference between them, although later interpreted in this way. Similarly, the scientific literature has shown that when a study has argued that a new medium is appropriate for educational contexts, it is not difficult to find another that shows the exact opposite (Hargrave, Simonson & Thompson, 1996). However, comparative studies continue to be performed in a truly significant number in the investigation of the new technologies as a didactic resource.

Of greater interest are the intramedium studies and ATI, molecular models but of much broader characteristics, which in our opinion surpass the defects—although they also have them, they are of less relevance and are adapted to the present characteristics of ICT—of evaluative research and comparative research. Already in the 1960s of the 20th century was the failure of comparative studies, which led many researchers to seek new lines of research that are more objective from a scientific perspective. In this case and because the media comparison had many gaps, we tried to do without such a fact to focus, in particular, on the attributes of each medium, i.e., to accurately identify their characteristics to determine their real possibilities of communication and interaction for the students and with the set of variables present in the didactic processes. Thus, as we pointed out, these two new models of research in educational technology have emerged.

In the case of the first, the intramedium studies, and following the proposal offered by Salomon (1981) and Clark (1985), investigations were designed in which they tried to measure all the variables other than the medium itself (presenting itself as a constant), so that it was possible to establish and compare different methods in which the same technology or medium was used. Based on Salomon's (1981) assertion that the effectiveness of each medium depends on the nature of the instruction, the question was not now to determine which medium is most effective, as was the case in comparative studies, but what is the most effective using that medium. The various experiences, in relation to ICT, that have been made based on the methodology proposed by this model (although they have not been too many in number, we must emphasize their quality) have been positive, and in some cases entirely satisfactory—Lehrer & Randle (1987), e.g., for informatics—and are helping to develop didactic strategies for the integration of new technologies as didactic resources. It therefore presupposes a model that we believe is certainly interesting and can be very useful for the study of the new realities of technological society, such as the phenomenon of social networks, Web 2.0 As a whole, the implementation of m-learning in education, MOOC courses, etc. Although it would take an independent study to accurately establish an assessment of the scientific results offered by these studies, we recommend its use as a line of work entirely superior to the traditional evaluative and comparative studies.

The same can be said of ATI research, which emerged in the 1970s as a new line of research that mainly participates in the constructivist paradigm. Until that moment most of the studies had been based on behavioristic elements, but following the outstanding psychopedagogical advances of that time, also began to work with the cognitive theories that define the learning like a process in which the student realizes an integration of the new knowledge about previous knowledge, and for which the media (as nowadays ICTs) are suitable: they are capable of producing external stimuli that develop cognitive processes that can be brought back to enhance learning. In this situation, however, different factors are taken into account, all of them considered to be extremely important: characteristics of the students, their abilities, previous knowledge, motivation, teaching methods, etc., highlighting the possibilities and types of interaction offered by the different means to enhance learning (Gómez Galán, 2016; Ponce, Pagán & Gómez-Galán, 2018).

In relation to this, research is beginning to be carried out based on this methodology, which has begun to be called ATI, for the use of these didactic resources and which was first defined by Cronbach & Snow (1977), although it was already outlined earlier by Parkhurst (1975). It was then developed by other authors such as Janicki & Peterson (1981), Kanfer & Ackerman (1989), Goska & Ackerman (1996), or Sternberg, Grigorenko, Ferrari & Clinkenbeard (1999). Examples of research from the last two decades, using this model for ICT and only as a sample button, are those of Kieft, Rijlaarsdam & Van den Bergh (2008), who investigated the importance of writing in learning, the ones of Seufert, Schütze & Brünken (2009), centered on memory and multimedia learning, those of Johnson, Lyons, Kopper, Johnsen, Lok & Cendan (2014), which analyzed the results and perspectives of students interacting with virtual patients in individual- and team-learning contexts, those of Yeh & Lin (2015), focusing on creativity through e-learning, the work of Lehmann, Goussios & Seufert (2016), centered on cognitive aspects of learning such as retention and memory or Lehmann, Goussios & Seufert, or the contribution of Park, Münzer, Seufert & Brünken (2016) to the study of spatial ability in multimedia contexts.

2.4 Conclusions

The results offered by these studies are that they establish that there are no better technologies or means or, *a priori*, with greater didactic possibilities within educational contexts, however sophisticated or ideal for teaching they are (even though they were created or designed for that end). It will depend on the individual characteristics of each student, or the group characteristics of a classroom, e.g., or the idiosyncrasy of a country, if we want to talk about a much larger population, or many other variables, that are adequate or not for a given teaching-learning process.

As we say, and due to the lack of resources still existing, it is certainly difficult to perform individually. However, the theoretical principles of these investigations, which from the practical point of view could be applied to groups of students, or large populations, previously analyzed to establish their characteristics and needs are extremely interesting. In this context, therefore, it is essential to establish with absolute precision the object of our research, since this is key to the methodological development to follow.

All methodology will always be dependent on what we want to investigate, why we want to investigate, and how, certainly, should be investigated to obtain the maximum information possible. It is essential, of course, in research processes in educational technology. This is just one example of what molecular models can do to encourage the integration of ICTs into educational processes.

3

Models in Educational Technology II: Global Models

José Gómez Galán

University of Extremadura, Spain, and Ana G. Méndez University, Puerto Rico-USA
E-mail: jgomez@unex.es; jogomez@uagm.edu

3.1 Introduction: Global Models for Overall ICT Integration

In our view, global models, and especially the selection of what we consider most appropriate, represent the main educational challenge that exists today. Facing the complex problem that we face only from a molecular perspective is insufficient. The importance of new technologies today is such that it is necessary to train students in these elements of such great impact and relevance in our society, if we really want to prepare citizens formed for the current life.

ICT education is needed, the presence of the same in the school as didactic auxiliaries or productive tools we consider problematic are already exceeded. It is essential to study, analyze, and critique the student, because nowadays it would be part of a comprehensive education adapted to the needs of a world marked by computer and telematic technologies, information, and communication processes. It is not the user of a product, in any way, as is sometimes considered, it is first and foremost a human being that is part of a society markedly technological. The integration of ICT, from this perspective, becomes an essential training challenge, in close connection with other social systems (political, economic, cultural, etc.). It is what we call a global dimension.

Educational research required for this is much more complex and, if not focused in an appropriate, indefinite way. How to face the study of this complicated object? The first thing that needs to be clarified, of course, is that learning is not the same as education. Education is the full formation of a person, with all that implies, between what would be essential to emphasize creating the need and the desire to learn. To educate means to integrate a person in a society, to make them imperative to their peers, to offer them and to give them a kind attitude towards life. In a more practical way, we can use what Burden & Williams propose (1997), i.e., that learning, a part of education, can only be educational when it adds value to the student's life. All education, therefore, needs learning although different parameters are closely linked. But in parallel, learning does not necessarily imply education.

The question, therefore, would not only be how to learn (which would lead us to an identical, indissoluble reflection of the former: how to teach—however abruptly separated, without apparent justification, in too many research processes— with ICT. It would also be, for what to learn and why to learn, in the context of a global education that allows the integral development of the person. In this sense we would no longer be in the field of technology in education, as discussed earlier. The key would be to understand the presence of these technologies and media today in our world, and what their real role in educational processes. Not only to be used as ancillary didactic resources, but to educate the student against the power of influence that they have in their lives.

On the other hand, we would have the very concept of technology in the field of educational sciences. Participating in the definition that Bunge (1985, p. 33) offers us as "the field of research, design and planning that uses scientific knowledge in order to control natural processes or processes, design artifacts or processes, or to conceive operations in a rational way" we could also understand educational technology as the conversion of education into technology, in a theoretical sense, thereby participating in the concept we previously alluded to, i.e., education technology. As it is a theoretical problem that is not the object of our work, and whose dialogue can be analyzed, in the field of the philosophy of science, by contributions such as Popper (1962, 1983), Mayr (1982), Feibleman (1983), or Bunge (1999), it may be said that it could also involve the creation of global models contextualized in this dimension.

3.2 The Field of Educational Theory

In short, theories that contemplate ICT in education can do so from multiple perspectives and evolve in parallel to their development: their presence in teaching-learning processes, their elements and interrelations, how to present information, their adequacy to existing methodologies (or the creation of specific ones for their use), the quality of materials, the appropriateness of when, how, and why to use their presence in the classroom, the theoretical structure that conceptualizes it, its implication in the educational processes above the instructive ones, the presence and importance of these technologies in our society and their influence in the education, etc. (Gómez Galán, 2017c). As we already pointed out, and we saw in the previous point, theories of communication and theories of learning are the ones that have had the greatest influence on our problem and therefore it is not possible to speak of a unified theory, consensual by the community (which is not surprising given the very large number of variables in the field of educational sciences, which is not the case in experimental or exact sciences, where there are an infinitely greater number of theories in the real scientific sense of the term).

The above are, together with *systems theory*, the fundamental pillars of the current conception of education technology. We will not, however, dwell on systems theory, as it is represented in the multidisciplinary methodology presented later (and especially in the global dimension). Logically, it tries to interpret the world as a set of interconnected systems (Bunge, 2010), which, as we have pointed out, would force the presence of ICTs in educational contexts (systems) because of their importance in other systems in which their protagonism is unquestionable. The interaction between them would also explain the complexity of the problem. Research from this perspective, and due to its lack of tradition, has not been adequately addressed by educational scientists and technologists. It would, however, be the best way to establish at present a comprehensive methodological model suitable for research in educational technology, flexible and open. Although we must also take into account that, possibly, we are not faced with a linear and identifiable process that allows such a precise concretion. It is more likely that parallel models are needed, and in dialogue with molecular approaches.

As we can see there are many possibilities in relation to global models in research in educational technology. We will not expand upon here, therefore, establishing a relation of possible models that, in many cases, would have difficult practical application and would be located, only, in the scope of the theory. We start from the fact that ICT cannot be considered only as didactic resource since its importance in society places them as fundamental elements of our world that it is necessary to know and analyze critically from an educational perspective, both by teachers and students. On the other hand, we have argued that there has to be a single, global strategy, which also includes the different molecular proposals that seek above all didactic objectives.

3.3 In Search of an Effective General Model: Theoretical Proposal

Would there be, therefore, a global model that would meet all these needs? The answer that we are going to offer, and that is affirmative, is related to what we have already presented before. It is appropriate to base ourselves on the analysis carried out for a long time in this sense, and in the world of techno-media convergence (Gómez Galán, 2003, 2007, 2011), in which ICTs today are intimately linked to the media (in reality they themselves are), to those that have been strengthened and extended, we can use research models that were really valid, were put into practice and, moreover, are perfectly adequate for the current methodological frameworks. And we refer, in particular, to those developed in the so-called media education. From it we can obtain a global model of ICT research in education from all perspectives previously analyzed.

When we talk about media education, we refer to the teaching of technologies and communication media so that students can reach the capacity to analyze, use, and even emit, in different ways the messages produced by them. It means in itself a need to educate for the current media society, where ICTs are today. Apparently everyone knows something about mass media. However, knowledge of the population is as a consumer, not as a user (Hart, 1991). Very few people are really capable of conducting a rigorous and critical analysis of media content. Almost all the basic characteristics of the communication of these media, their codes, messages, and methods of influence are unknown. This leads, e.g., to the fact that the consumption of their products is in most cases absolutely irrational. Even that leads to different addictions, dependencies, etc.

The population, in a global way, is more demanding for any other type of products than for those elaborated by mass media. And the main reason, among others, is that society has not received adequate training in them, neither as consumers nor, much less, as users. And in this context the role of the teacher as a guide is fundamental. The students must be able to carry out, in a critical sense, the analysis of the information that comes to them through ICT, powerful media whose influence is undeniable (Gómez Galán, 2003). As stated by Jenkins, Clinton, Purushotma, Robison & Weigel (2006) it is necessary that today's students acquire new skills in the new media culture that envelops our society.

Of course we are talking about a challenge of education in relation to current social needs. Any educational process that pursues the training of active and critical citizens must now have an impact on teaching the correct use of ICT. In addition, analyzing the significant means giving personal freedom to each individual, which is an essential function in any democracy. As a whole, critical literacy of the population can create a global awareness of action that makes people present their own opinions and thus avoid being easily manipulated by powerful media groups (Masterman, 1990, 2001; Buckingham, 2003). It is what it would mean to create, in essence, an active democracy. Educational research in ICT should have a special impact on this. These elements of our society, as we defend, are much more than mere tools that can be used as didactic resources. They constitute one of the bases of the society of the 21st century.

Because of all of this, we defend to contemplate the state of the situation in a global way, not essentially pedagogical, since there have been multiple perspectives from other fields (sociological, psychological, political, semiotic, journalistic, etc.) that have made inroads into the educational implications of the media and new technologies, and which in the end had greater influence and acceptance than those made by specialists in education sciences. But the pedagogical dimension must always be present, otherwise the research processes would incur very dangerous reductions due to their consequences. For example, the paternalistic sense of media education that developed in the 1940s was more responsive to social and political than pedagogical pretensions, and had a great influence on later studies.

At the beginning of the 1980s the research was focused on studies of an artistic and aesthetic nature in which attempts were made to establish strategies for the student to acquire the capacity to establish quality criteria. It would be only after that decade that a new conception arises in which critical analysis from a representational and political perspective (media products as representations to achieve certain ends) replaces aesthetics (Masterman, 1990), and begins to be considered seriously, from the contributions of Hall (1977), the power of the mass media to spread ideologies. Research in media education will began to extend from the Anglo-Saxon countries, from where it was conceptually and methodologically born, to the rest of Western educational systems, but a long way from the former.

At present, and in a particular way, we defend the use of research models in ICT research processes in educational contexts, both as didactic resources and as the object of study (overcoming the division in both dimensions), since new technologies are today at the service of the powerful mass media; and are, of course, means. And we have done so from a theoretical perspective as well as developing a field research model, which would summarize in practice all that we have exposed in this chapter (Gómez Galán, 1999, 2003, 2014c; Gómez Galán & Mateos, 2002).

3.4 Conclusions

Naturally the methodological frameworks maybe in line with the specific objectives pursued, and the educational researcher may rely on those that they consider best suited to their needs. Although we have proposed a global model, it would also be possible to use molecular models from this perspective. For example, on the didactic and curricular approach of teaching for the media we find some methodological proposals of interest, even in the 1960s—such as those of Hall and Whannel (1964)—that in due time made them abandon the paternalistic and protective attitude based on the condemnation of the media, to promote the debate on the value or importance of media messages. This gave rise to representational models that basically advocated, as we are exposing, the need to study these important elements of our society, their capacity to be creators and mediators of social knowledge, so that future citizens of any society need to know how they represent reality, how they do it (techniques they use) and what ideology lies behind those representations (hence the name of these models, the ability of the media to create reality, but which are not, of course, the same reality), i.e., why they are created and extended. The best examples of this line of research are, among others, the proposals and reviews presented by specialists such as Masterman (1990, 2001), Christ & Potter (1998), Hobbs (2004), Hobbs & Jensen (2009), Potter (2010, 2013), Pangrazio (2016) or Rasi, Vuojärvi & Ruokamo (2019). UNESCO itself in 1984 had already been aware of the importance of studies in this regard. In our particular case, we offered a complex model more than a decade ago (Gómez Galán, 2003) perfectly applicable, as we have recently defended, today (Gómez Galán, 2015b).

Media education (which today includes ICT, which we consider unified for all the reasons we have presented), from this new paradigm of research, plays a decisive role in the training of future citizens (in some way treated as a formative and/or cultural complement). This new perspective is so significant that it could even modify the very consideration of the nature of the educational curriculum. And the challenge today is indisputable, increasingly underlined by the great development of ICT in recent years, and the convergence to the digital world has led to communication processes.

Along with a second youth of traditional media (also driven by their development from digital technologies), we are today witnessing the explosion of Internet and social network consumption, multimedia technologies, the extraordinary development of the power and possibilities of computing and telematics, mass and cloud storage systems, video games, virtual and augmented reality, the unlimited possibilities of tablets, etc., as a preamble to future technologies, still under development, which allow and will allow the media to offer information with a quality (and quantity) as never before achieved. With all that this will imply, of course, in terms of its influence on society and of course on education. There will be only one answer to all this: ICT curriculum integration must be carried out in a serious, precise, and rigorous manner. This is the main objective of the lines of research that, in educational technology, will be a priority in the future.

4

The Complexity of Education in a New Social Scenario

Omar A. Ponce

Ana G. Méndez University, Puerto Rico-USA
E-mail: um_oponce@uagm.edu

4.1 Introduction

The link between research and the improvement of society is an ideal that guides contemporary scientific research. In education, identifying what works and does not work in schools has put much emphasis on the discussion of causal relationships studies (O'Connell & Gray, 2011; Biesta, 2015). The study of causal relationships is associated with the improvement of educational quality and school effectiveness. Research on school effectiveness and educational quality is one of the most researched topics in education. This is why linear studies of cognitive outcomes dominate (Galán, Ruiz-Corbella & Sánchez-Mellado, 2014). Educational research also requires studies of causal relationships to better link education theory with practice and the formulation of effective educational policies (Radford, 2006; Cheung & Slavin, 2016). The study of causal relationships in education is an issue that arises whenever the effectiveness of teaching methods, curricula, programs, or schools is discussed in promoting student learning. Effectiveness in the study of causal relationships in education is an issue that is discussed and is intensely debated among educational researchers. The study of causal relationships is the greatest challenge of scientific effectiveness that has faced educational research throughout its history. It has therefore been difficult in the field of education to anticipate situations and predict outcomes. This, in part, has implied that educational research is perceived as an elusive and imprecise science. This chapter discusses the challenges of investigating causal relationships in the complexity of education. Recommendations are made to improve research on causal relationships and the need for a new paradigm for educational research.

4.2 The Challenge of Researching Causal Relationships in Education

The concept of causal relations reaches the field of educational research in the early 20th century (Johnson, 2001). The concept seems to be generated in the physical sciences to explain the chain reaction between phenomena or events. For example, water boils at 100C. As the intensity of the heat increases, it exerts an effect on the water. Once this relationship is known, this knowledge can be used to purify the water or change its state to gas and thus provoke other reactions and other events. Knowing that a relationship exists between events or situations, and being able to measure accurately when the relationship occurs, allows predicting when it will occur, preventing its results, managing them if they cannot be avoided, or manipulate them if desired. The study of causal relationships allows humans to conquer and control their environment. To establish that there is a causal relationship between two events, there must be three conditions: (a) The relationship between the two events actually occurs. In other words, the relationship between events is not a random occurrence or an accident. (b) The manifestation between events manifests itself in a temporal way. This means that there are always conditions that provoke the "cause" and cause it to manifest first and then a reaction that detonates the second event or the "effect" occurs. (c) The relationship is not the product of other circumstantial events or from strange or external variables. Therefore, to explain a causal relationship, researchers must be able to generate an evidenced explanation of the relationship (i.e., how it occurs and why), but also to be able to explain the events, dynamics, or circumstances that cause the cause (i.e., what causes it and why) and the dynamics, circumstances, or events that constitute the effect (what it is and why) (Miles & Huberman, 1994).

The field of education is structured, based on multiple causal relationships. The most common of these is the relationship of teaching and learning. Teachers use a series of teaching strategies that will produce learning in students. The rationale is that teachers create environmental conditions and the right climate in the classroom and strategies cause students diverse ways of knowing and understanding knowledge, and applying it to practice. With this same logic, curricula and academic programs are designed and organized where the premise is that certain activities and educational experiences are presented to students in certain sequences to produce the desired development and learning. In this way, educational systems are organized and operate based on educational objectives, academic policies, and administrative structures that delineate the direction of teaching work in schools, establish the functioning of the components and define the causal relationships between their constituents.

4.3 Educational Practices and Results

In schools, multiple relationships occur in their management to produce student learning. There are relationships between nonanimated objects and people. For example, the relationship between a teacher's teaching methods and student learning, the student's relationship with his or her curriculum, or school policies. There are also multiple human relationships among students, teachers, and other constituents of schools. For example, a child whose parents go through a divorce process and brings that drama to their classroom. This situation may cause other reactions in classmates or teachers. This is considered a level of complexity difficult to study (Miles & Huberman, 1994; Radford, 2006; Galán, Ruiz-Corbella & Sánchez Mellado, 2014). Human beings are not always rational and predictable, but act by emotions, values, and beliefs because they interpret the situations that occur to them. Therefore, in the field of education human relations are not always linear, but multidimensional, conjunctural, and temporal because they change over time (Miles and Huberman, 1994; Galán, Ruiz-Corbella & Sánchez Mellado, 2014). Many of the variables in the multiple human relations of education cannot be reconciled (Johnson, 2001), nor the complexity of relations simplified or reduced to isolated or decontextualized variables of school dynamics to study (Radford, 2006; Galán, Ruiz-Corbella & Sánchez Mellado, 2014).

The study of causal relationships in the complexity of education has generated intense discussions and polemics about the relevance and superiority of research designs on others to study education (Johnson, 2001; Shavelson, 2015; Kaplan, 2015; Rowe & Oltmann, 2016). It has also generated among scholars and researchers the idea that the concept of causality in the natural sciences does not apply to the study of human beings (Miles & Huberman, 1994; Radford, 2006; Galán, Ruiz-Corbella & Sánchez Mellado, 2014).

4.4 Designs of Educational Research in the Study of Causal Relations

It is recognized that educational research has made great strides in the sophistication of its research designs and in its efforts to solve the problems of education (O'Connell & Gray, 2011; Galán, Ruiz-Corbella & Sánchez Mellado, 2014). It is also recognized that to solve contemporary problems of educational effectiveness, quantitative, qualitative, and mixed methods of research is needed. Each model provides to confidently address diverse questions of education and its complexity (Shavelson, 2015; Ponce, 2016). Not recognizing this is trying to erase a scientific history (Galán, Ruiz-Corbella & Sánchez Mellado, 2014). To study causal relationships, educational researchers resort to a range of experimental and nonexperimental research designs (Johnson, 2001; Redford, 2006; Cheung & Slavin, 2016). Let us examine the strengths and challenges of educational research methods in the study of causal relationships and the complexity of education.

The quantitative research designs used in the study of causal relationships are experimental design, quasi-experimental design, design of cause comparison (ex post facto), and correlation design (Ponce & Pagán-Maldonado, 2016; Ponce, 2016). Causal comparison and correlation designs are common in education because there are many variables that cannot be controlled at the time of study. Researchers have to resort to these designs to study them, as they are manifested in schools (Johnson, 2001; Ponce, 2016; Ponce & Pagán-Maldonado, 2016). The model of causal relationships has focused on identifying possible relationships with correlation studies that are then sought to validate with experimental studies (Redford, 2006; Galán, Ruiz-Corbella & Sánchez-Mellado, 2014). For some scholars and researchers, designs with random samples, such as the experimental design, are considered more powerful to study causal relationships in the field of education because they allow control of study variables. Random selection of participants reduces bias in the study of groups and minimizes the possibility of competing explanations to observe causality (O'Connell & Gray, 2011; Shavelson, 2015). The use of statistics facilitates the creation of mathematical models of these relationships (O'Connell & Gray, 2011), which can then be accurately controlled and measured through experimentation, and this helps to minimize errors in the interpretation of causal relationships which are studied (Shavelson, 2015; Kaplan, 2015).

The major criticism of experimental and nonquantitative experimental designs in education is that they are reductionist and deductive research approaches. This makes them weak and narrow minded to capture the complexity of education (Radford, 2006; Feinberg, 2012; Galán, Ruiz-Corbella & Sánchez Mellado, 2014; Rowe & Oltmann, 2016; Roni, Merga & Morris, 2020). From epistemological and ontological perspectives, experimental methods are based on a positivist and postpositivist philosophy, where knowledge is understood as an entity separate from people and reality as an objective entity. For example, the knowledge teachers teach is independent of students. If the student does not learn then the cause must be in teaching methods and teachers. This type of interpretation can distort facts and negatively affect education by comparing schools because it does not consider or capture the complexity of education (Rowe & Oltmann, 2016) nor the implicit values of the situation (Feinberg, 2012).

4.5 Conclusions

There is no doubt that experimental and nonexperimental quantitative methods can help to understand the complexity of causal relationships in the field of education (Gavira, 2015; Rowe & Oltmann, 2016; Gore, 2017). Educational researchers should exercise extreme caution not to confuse themselves with statistical complexity to distort the interpretation they make of the reality they study and the evidence they generate (Rowe & Oltmann, 2016). In the future, experimental and nonexperimental quantitative research in the culture of evidence-based research can be strengthened by adopting the following research practices and strategies:

(a) Ensuring the quality of the sample that is selected (O'Connell & Gray, 2011). Use probability samples whenever possible (Cheung & Slavin, 2016). (b) If possible, use the statistical models to represent theoretically the complexity of the educational phenomena studied (O'Connell & Gray, 2011). (c) Make more use of standardized instruments than instruments developed for particular and small-scale studies. Standardized instruments produce more reliable and comparable data between schools and groups than instruments developed for the specific purposes of a particular study. (d) In program evaluation studies, where the audience is both politicians and educators, large samples should be sought to produce convincing and outreach evidence. (e) Design studies based on theories that confirm or reject data or evidence (O'Connell and Gray, 2011). This is summarized in studies that are carefully designed to produce reliable evidence (O'Connell & Gray, 2011; Cheung & Slavin, 2016).

In the 21st century, qualitative research is projected as a research model of social criticism that seeks to contribute to the solution of the world's social problems (Denzin & Lincoln, 2011; Flick, 2016; Lester & Nusbaum, 2018; Muthanna, 2019). Three research statements are identified in this positioning to contribute to a better society (Flick, 2016). Qualitative research needs to be critical in exposing and explaining the social and political issues under study. Qualitative research needs to identify vulnerable groups in society, define the problems that affect them, and analyze how institutions handle these problems. It also needs to make recommendations on how these problems can be solved. In this way qualitative research will remain relevant and useful to society.

5

The Scientific Study of Education in the Technological 21st Century

Omar Ponce

Ana G. Méndez University, Puerto Rico-USA
E-mail: um_oponce@uagm.edu

5.1 Introduction: Between Science and Politics

In the field of education research needs to be questioned in the context of the evidence-based research movement, the prevalence of experimental research as the only way of investigating causal relationships and the fascination with mixed methods. It is also necessary to be critical of the methods of qualitative research that are used as scientific research in education.

The objective of questioning the practices under the name of scientific research in the field of education, which includes qualitative research, is not to discontinue some forms of research to give way to others. It is a matter of questioning premises to better understand the methods, the data that are generated, to produce better reports, and politically to eliminate the marginalization of the qualitative research of the participation of governmental funds.

In the field of education, causal relationships have been successfully studied with field observations studies (O'Connell & Gray, 2011), with case studies (Flyvbjerg, 2011), and with ethnographic studies (Ponce, 2014). Qualitatively studying causal relationships involves deeply describing events or situations that are causal, describing in depth the effects, connecting and describing the processes of causality, and explaining the institutional context where the causal relationship with temporal elements manifests (Miles & Huberman, 1994). The major strength of qualitative research is its flexibility to capture the complexity of education and its multiple relationships

(Cooley, 2013; Galán, Ruiz-Corbella & Sánchez Melado, 2014; Ponce, 2014; Ponce & Pagán-Maldonado, 2016; Ponce, Gómez Galán, & Pagán, 2019). The greatest criticism has been the quality of evidence generated by methodological shortcomings. According to Cooley (2013), qualitative research in education has given simple answers to complex questions. In a culture of evidence, qualitative research can be strengthened by adopting the following strategies (Hammersley, 2008).

5.2 Refining Qualitative Research Designs

Qualitative researchers have never developed a robust stance on the methodological deficiencies noted in qualitative research: lack of precision, ambiguity in the study of causal relationships, and lack of generalization. The omission of these criticisms, arguing that they come from the lack of knowledge of the qualitative model, has been negative for qualitative research because they were never used to strengthen the model. More robust positions on these methodological shortcomings are needed.

The qualitative research failed in its original intention to understand the human behavior in schools, from the viewpoint or the reality of the people. The selection of study groups (samples) has been based on ideological perspectives or from the perspective of marginalized or political groups. This has exposed the lack of information about some social groups. The selection of study groups warrants the same care and rigor as other models of educational research in their selection.

Qualitative research that seeks to study the understanding of the participants, or the processes through which constituents pass in educational institutions, structurally carries the possibility of misinterpretation. Qualitative research has been criticized for being inaccurate, for methodological errors, and lack of rigor. For example, there are studies of educational processes that relied on interviews when observations were to be made by the participant. In many qualitative studies it is assumed that the participant tells the truth when interviewed or that the information is valid. Therefore, no other forms of corroboration of the data are sought. Another example is that multiple forms of information are collected through documents, interviews, or observations, but these are not corroborated either. Accuracy of data and interpretation needs to be strengthened.

The proliferation of methodologies and philosophies of qualitative research is celebrated, when it should be repudiated. This produced a methodological fragmentation that pushed qualitative research away from its original intention in the field of education to understand student and educational processes.

The studies of social criticism, the postmodernists, and the reconstructionists, where the emphasis is to expose the voices of the oppressed, distanced qualitative investigation from its educational intention.

Schofield (2007) argues that generalization or external validity is a relegated issue in qualitative research. Identified two reasons in this regard are (1) the little importance that the subject merits for some researchers and (2) because the generalization is considered something unattainable. In the field of educational research, the generalization of qualitative studies has become an important topic that merits attention. Qualitative research played a prominent role in research and educational evaluation. Contemporary qualitative educational research is not limited to the study of exotic cultures, but as a tool to address the real problems of education and society. The use of qualitative research as part of the evaluation of academic programs on large-scale, federally funded projects reaffirmed this need. Many educational researchers express an interest in transcending the traditional practice of qualitative research to describe the phenomenon they study to adhere to the possibility of applying their findings to other educational scenarios they studied. The issue of generalization is recurrent in the allocation of funds for educational research. Working the issue of the external validity of qualitative studies in education merits a rethinking of the traditional concepts of generalization that emerged from quantitative research. Although there are works in this direction, these are not conclusive. These are the following consensus on the generalization of qualitative research, or any social research: It cannot be the search for laws of universal application. It does not imply that the data of a particular situation can inform another similar situation. The deep description of the phenomenon under study is fundamental to be able to identify the similarities and the differences that allow that the data of a situation can help to inform another. In the field of education, educational processes and institutions are studied. So, generalizing in this context entails three types of approaches: studying what is, studying what can be, and studying what could have been.

In the 21st century, qualitative research needs to be accessible and to establish bridges of communication with the political world and with educators. Qualitative research must address the problems of education systems and generate rich and detailed information about these problems, and their institutional and cultural contexts, to respond to the needs of educational policies and the teaching strategies needed by educators. Qualitative research needs to generate practical solutions for educators and politicians in the fiscal reality in which many educational systems live. Practical research becomes a useful research. In the 21st century, qualitative research data should promote and facilitate communication channels between politicians, educators, and administrators about what happens in schools (Sallee & Flood, 2012; Cooley, 2013). Understanding the problems of schools in their contexts and cultures helps to improve educational systems (Galán, Ruiz-Corbella & Sánchez Melado, 2014; Ponce, 2014; Ponce & Pagán-Maldonado, 2016; Ponce, Gómez Galán & Pagán, 2019) and to have more responsive educational policies (Sallee & Flood, 2012).

5.3 Objectives of the Mixed-Methods Research

In the 21st century, mixed-method research is understood as a way of knowing from multiple perspectives (Creswell, 2016). The strength of mixed methods research in education is the possibility of combining quantitative and qualitative methods in the study of a problem and the scope it has for capturing the complexity of education (Creswell & Garrett, 2008; Ponce, 2014; Mertens, 2015, Ponce, 2016; Ponce & Pagán-Maldonado, 2016). This possibility includes the study of causal relations (Ponce & Pagán-Maldonado, 2015; Ponce, 2016; Ponce & Pagán-Maldonado, 2016; Ponce, Gómez Galán & Pagán, 2019). According to Ponce & Pagán-Maldonado (2015), the complexity of education emanates from the phenomena investigated (e.g., teaching and learning processes), educational contexts (e.g., school culture and policies), and of its administrative structures with multiple levels of functioning (Ponce, 2016). This same attribute is recognized for the research of mixed methods in other academic disciplines (i.e., Mertens, 2015; Jokonya, 2016).

Mixed methods provide the possibility of measuring the phenomena that occur in education and in the context, culture and policies of the schools where they are manifested. In educational research, triangulation and complement designs have been successfully used in parallel phases to measure causal relationships related to teaching and learning processes and to environmental factors and conditions with work stress (Ponce and Pagán-Maldonado, 2015). Multi-sample and multilevel designs provide a way to approach the diversity of educational dynamics of interrelationships of people or groups or that are products of the multiple structural levels of educational systems (Ponce, 2011; Ponce, 2014; Ponce & Pagán-Maldonado, 2015; Ponce, 2016; Ponce & Pagán-Maldonado, 2016).

The major criticism of mixed methods research is its complexity because it involves combining or integrating quantitative and qualitative research methods into the research of a problem to capture it in all its extent, scope, and complexity. This implies mastery of quantitative and qualitative research methods. Mixed-method research consumes time in education (Ponce & Pagán-Maldonado, 2015) and in other academic disciplines (i.e., Jokonya, 2016). The complexity of mixed methods has raised the question of whether this research model turns out to be more appropriate for teams of researchers than for individual researchers (Creswell, 2008).

In the 21st century, mixed-method research faces several challenges (Creswell, 2016): The need to develop new research designs that capture the emerging complexity that reflects the problems of humanity. Society is becoming more and more complex and this complexity challenges the existing mixed designs. Many of the current designs are provided for projects carried out by individual researchers. This is to achieve a paradigmatic vision of greater openness and creativity so that the investigation of mixed methods can capture the plurality existing in society. It is imperative to develop the concept of integrating research approaches to address those problems where the researcher needs to transcend combining quantitative and qualitative approaches.

5.4 Technology in Educational Research: Analysis in the Context of Causal Relationships

In the natural sciences, technology is recognized as a core component in the development and effectiveness of research on causal relationships. By technology, we mean the use of instruments, devices, and technological equipment that have allowed to identify and study causal relations imperceptible to the human eye. As a result, research in the natural sciences has been effective in studying causal relationships, in measuring the direction and intensity of these, and in anticipating, predicting, or controlling and preventing outcomes of situations or events where these relationships' causes and effects are manifested. As a result, research in the natural sciences (biology, chemistry, and physics) has been very effective and useful in applied disciplines such as medicine (biology), agriculture (chemistry), and engineering (physics). Educational research in a technological age is an emerging issue (Spector, Johnson & Young, 2014; Bakir, 2016; Knezek, Christensen & Furuta, 2019).

In educational research, technology has been used in various ways to improve the accuracy of data and generate a better understanding of education. For example, video recorders and photographs are used to collect data, to document behaviors, events and dynamics of education, and to be able to study them repeatedly after they have been documented (Torre & Murphy, 2015; Nebel, Schneider & Rey, 2016).

Computer programs have been developed to analyze quantitative, qualitative, and mixed data (Creswell, 2016) that have allowed greater precision in their interpretation and the possibility of exploring causal relationships imperceptible to the human eye or to logical thinking. Programs and databases have been generated on institutions that allow them to analyze their operation, use information to make managerial decisions, and determine their effectiveness (Orlikowski & Baroudi, 1990). There are computer programs to analyze electronic documents and facilitate documentary research of hundreds of documents systematically and in less time than the traditional method of identifying and reading documents (Chang, Long & Hui, 2013b).

In the investigation of causal relations we have been using virtual simulations where different statistical models are applied to test and measure these without having to enter the schools. The findings of these simulations can be tested with experimental methods to validate them (Galán, Ruiz-Corbella & Sánchez Mellado, 2014). Diversity of electronic equipment has also been used to study the functioning of the brain and how the human being learns (Ellis, 2005). Brain research has generated data that raise questions about the wisdom of existing theories about learning, school organization, and educational practices (Ellis, 2005).

Technology in educational research is an important tool for continuing to enter the complexity of education in the study of causal relationships. Despite these advances, there are educators and educational researchers who think that causal relationships research is an unachievable ideal in the field of education. In the view of these academics and researchers, educational research needs to adopt a research paradigm on complexity and thus advance research on the effectiveness of education.

5.5 Capturing the Complexity of Education in the 21st Century

The dominant position in literature is that the research of causal relationships is an important issue in education because it impacts the effectiveness of schools. It is also understood that educational research has managed to advance the subject of the study of causal relations because it has refined its research methods. Most of the recommendations documented in this chapter are aimed at strengthening methods of educational research. The dominant view is that educational research must be pragmatic and respond to the needs of education. Technology in educational research is a tool to obtain more precise data and to study the causal relations. In spite of these advances, the argument arises that the investigation of causal relations seems to be an ideal unattainable by the complexity of education. To advance the issue of educational effectiveness, educational research must adopt the paradigm of institutional complexity that allows it to examine causal relationships in a realistic and compatible way with the nature of schools and their purposes (Galán, Ruiz- Corbella & Sánchez Mellado, 2014).

The critique of causal research in the field of education is that it does not capture the complexity of school reality. Much of the causal research is part of the results of standardized tests that seek to connect educational activities (i.e., teaching techniques, curricula, programs) with results (Biesta, 2015). The premise is that learning outcomes can be explained in causal, linear, and logical relationships between activities and their products or outcomes. Causal research approaches the educational phenomenon by isolating the study variables and extracting them from the context and the educational reality. In doing so, educational processes are obviated, other important causal variables can be hidden, and school reality can be distorted (Radford, 2006; Galán, Ruiz-Corbella & Sánchez-Mellado, 2014). For example, explaining the achievement of educational standards excludes the issue of diversity that occurs in schools. It is for this reason that causal research alone cannot be considered a complete approach to the study of the effectiveness of schools. The complexity of education is an issue that is discussed in relation to the effectiveness of educational research (Radford, 2006; Galán, Ruiz-Corbella & Sánchez Mellado, 2014; Ponce & Pagán- Maldonado, 2015, 2016; Ponce, 2016; Trimmer, 2016).

Authors such as Radford (2006) and Galán, Ruiz-Corbella & Sánchez Melado (2014) argue that educational research must be approached from the paradigm of complexity and chaos. The paradigm of complexity and chaos originates in the quantum sciences in the 19th century, was used during the second half of the 20th century in social research and began to be seen in the field of education from the 1990s, approximately (Galán, Ruiz-Corbella & Sánchez Mellado, 2014). The complexity theory postulates that a system is complex when its operation is focused on multiple relations and with multiple components, interacting simultaneously (Galán, Ruiz-Corbella & Sánchez-Mellado, 2014). Problems in complex systems are chaotic and are characterized by the following: (a) They create uncertainty, a sense of urgency and the need to solve them in relation to the time factor or "we have to solve them because the deadline is approaching," (b) there is no perfect knowledge for its solution or there is no single correct way to solve them because the solution is largely related to how the dynamics are interpreted and understood, (c) sometimes the person or persons who try to solve them complicate them in turn, and (d) the available institutional policies do not provide alternatives to manage the problem so that it is not repeated in the future. Traditional methodologies of using a large amount of data and statistical programs to analyze information are not sufficient for decision-making and solving them. Examples of problems and dynamics in complex systems maybe inequality, access to services, or terrorism

Complex systems always operate in a dichotomy between order and disorder. Order is the desired functioning of the system and disorder always constitutes an opportunity for improvement and innovation to develop it if it decomposes, deviates, or aspires to higher levels of effectiveness. Complex systems show effects and products that are not always equitable to the "cause" because their relationships are not always linear, orderly, and logical. This has been evident in the study of the organization of clusters of certain species of animals, such as ants, where teamwork is the axis of operation. For example, in the presence of danger, a group member may emerge as a leader and undergo an internal reorganization of the system and group. The analogy of this in school maybe that the academic failure of a student is not the product of the school but of the student or home from which it comes. Another factor affecting complex systems is the variability of time. This implies that certain operations and certain tasks must occur on certain dates for the effective and desired functioning that produces stability. In the study of the effectiveness of schools, the variable time has been studied very little (Galán, Ruiz-Corbella & Sánchez Mellado, 2014).

5.6 Conclusions

In the pursuit of educational efficiency, educational research has to approach schools as complex systems. It is necessary to understand that the product or the results of complex systems such as school do not come from a single set of causes but from the sum of many multidimensional relations that are not always linear. These multidimensional relationships can only be understood if they are studied in the context of the organization. Therefore, the enhancement of these interactions and relationships can only be achieved in the context of organizational structures. This implies a redefinition of the research approach of school effectiveness that recognizes the functioning and complexity of the reality of schools. This does not rule out linear research on causes and effects that explain the results of standardized tests, but provides the space for those who wish to investigate education as a tool for student development to do so (Biesta, 2015). In the 21st century, educational research needs a paradigm that organizes it to address the complexity of education and educational systems (Radford, 2006; Galán, Ruiz-Corbella & Sánchez Mellado, 2014; Ponce, 2016 Gilbert, 2019). Being able to understand the functioning and effect or results of the multiple relati; onships that occur in schools in the context of their organizational, administrative, and policy structures should facilitate improving the efficiency of schools and education systems. This should facilitate improving the practice of education and educational policies in the context of the complexity of education systems.

6

Education and Knowledge

Nellie Pagán Maldonado

Ana G. Méndez University, Puerto Rico-USA
E-mail: npaganm@uagm.edu

6.1 Introduction

Since the time of the Copernican revolution, scientific research has been understood as the search for knowledge. Science means knowledge (Thompson, 2012). This knowledge has to increase the effectiveness of the profession in the following way: (a) generate theories that explain the practice of the profession, (b) validate these practices, and (c) produce laws that precisely guide the profession toward its achievements. In this way, professionals will be able to intervene, control, or predict the events or their results. In this third expectation of science lies the challenge of the effectiveness of educational research. At the time of this writing, educational research has not found and has not been able to generate that universal teaching method that guarantees learning or that mechanism which makes it possible for all students to complete their studies. Between the 1980s and 2000s, there was a great deal of controversy among educational researchers about the nature of knowledge (Paul, 2005). This debate seems to have continued for a long time even though it does not lead to any solution (Fernández-Ramírez, 2014).

Contemporary educational researchers are interested in understanding what knowledge is generated from education and what field of knowledge it is (Green, 2010). Educational research needs to be paradigmatically aligned in the continuum between the objectivity and subjectivity of education. There are many educational research philosophies that answer different questions about social reality (Marley & Levin, 2011). The diversity of research philosophies brings a wide range of interpretations about the knowledge and nature of education (Cumming, 2012). It is for this reason that educational research needs to start from a clear statement of values and beliefs about what is considered true in education systems. The understanding of epistemology and ontology in research helps to produce studies with greater precision because they are customized to the situation (Cumming, 2012). In this sense, it is important to be clear how the field of education originates and what is pursued to keep developments focused on where educational research should be directed (Green, 2010).

At the beginning of the 21st century, attention should no longer be to the nature of knowledge, but rather to identify what knowledge is needed in the field of education to make it effective (Green, 2010). To specify the knowledge needed in education can be a first step in the search for this paradigmatic consensus on educational research. In the past, neither the philosophical nor the methodological approaches to education and its research have produced the paradigmatic alignment needed at the beginning of the 21st century. This chapter discusses the salient points on the subject of knowledge that can be generated from the field of education and the methodological considerations that must be kept in mind in the search for this scientific ideal. The subject of knowledge can be a meeting point where the paradigmatic diversity that currently exists in educational research can be aligned.

6.2 What Is "Knowledge"?

The word "knowledge" means many things: comprehension, understanding, discernment, judgment, information, and erudition. Knowledge is a product of the mind because it is a construction or a mental exercise. Knowledge is "knowing the essence" of the phenomena we investigate (Ponce, 2014); describes and corresponds to the phenomenon under study (Scott, 2014); and is generated in a relationship between the researcher and the phenomenon under investigation. The research phenomenon may have tangible and real properties (e.g., classroom temperature) or intangible and perceived (e.g., whether the room temperature is pleasant or not). In educational research the knowledge generated by a researcher consists of the formation of an idea or an image of the object under study (understanding). This knowledge is real when it corresponds to the represented object or is investigated because it describes or explains it in observable manifestations. In the scientific culture of contemporary research, researchers are expected to be able to describe or explain the phenomena they study. Describing means being able to define the educational phenomena that are investigated, identifying its components, or identifying how they originate. Explaining means being able to describe how the phenomenon works and interacts with other components or aspects of education. The authenticity of this knowledge lies in the agreement between the mental image formed by the researcher with the knowledge generated and the object or phenomenon investigated. For educational researchers, the best way to achieve this knowledge and validate it is to place themselves into educational scenarios (field research) where the problems to be investigated are (context) (Ponce, 2014). This was one of the reasons that led to the abandonment of laboratory experiments in education to make them in real educational environments (Condliffe & Shulman, 1999). This means that the act of investigating education does not only consist of describing or explaining the phenomena that are studied, but also includes the exercise of establishing and corroborating the correspondence between the knowledge generated by the research and the phenomenon being studied. In this way one can argue about the authenticity of the findings and the contribution that is made to the field of knowledge (Marley & Levin, 2011).

The great challenge of validating the knowledge of the subject under study is to be able to identify the observable manifestations of the educational phenomenon under investigation. For example, Jean Piaget, the French psychologist, developed a theory of how learning occurs in children. Piaget's explanation is that children go through a series of mental stages, from acquiring information until they can use it with mastery. The mental stages of cognitive development theory cannot be directly observed in the brain (physical reality), but this managed to link them to observable behaviors that facilitate appreciation of the child's learning in the light of his theory (social reality). In the 21st century, educational researchers should put equal emphasis on generating knowledge and validating it.

6.3 A Wide Discussion on Knowledge

The vision of knowledge that emerges with the idea of scientific research focuses on the possibility of objectively knowing an entity. This positioning focuses on the notion of a method that allows us to study and know the entity objectively, by excluding the researcher's prejudices (Thompson, 2012). In this vision or view of knowledge, physical qualities are attributed to the phenomena under study. The entities of investigation exist around us, they are real, they are objective, they are palpable, and they are independent of us. Even if we are not, they will remain present in social reality. Adopting this research posture involves assuming a realistic research philosophy. To investigate from this perspective of knowledge implies looking for the absolutes or facts of the phenomena studied and to generate evidence that allows corroborating these (Fernández-Ramírez, 2014).

From the 1970s to the present, the realistic and absolutist view of knowledge has been questioned by a group that supports the thesis that knowledge is a mental-social construct. Knowledge is a consensual understanding that individuals make of the social reality of life they share. Social reality consists of the dynamics of beings in society, among them and with their environment. Knowledge cannot be judged outside the logic of reason and the cultural-social context where individuals interact and is captured with language (Thompson, 2012). To adhere to this vision of knowledge is to support a relativistic and constructivist philosophy of knowledge. From this perspective, research cannot always generate physical evidence of phenomena because they are social and symbolic constructions of cultures. In other words, social interactions do not have physical properties that can be measured or that exist beyond the meaning that people assign to them (Fernández-Ramírez, 2014).

Since the 1980s, the insertion of neoliberal philosophies into the field of education, advances in information technology, and the need for accountability has transformed the vision of knowledge for information. The new economy of information and information technology links information with its usefulness. The usefulness of knowledge/information

lies in its potential to solve the problems of humanity. The information has economic value and development. For example, the country that discovers the cancer cure can control that market by patenting that knowledge and putting it on sale. Organizations can store information about their operation and maximize their operation through computer programs that make it easier to compare their operation with previous years or with other organizations. In this way, information in the binary form of computer programs allows organizations to store information, not repeat past mistakes, and explore unsuspected relationships through artificial intelligence provided by computers. This has given way to a new approach on the creation, storage, and use of information (Yong Chung & Yoon, 2015).

In the 21st century, the debate on social knowledge continues to exist (Fernández-Ramírez, 2014) in the field of education, but it is acknowledged that it does not advance the need for development that has educational research (Green, 2010; Fernández-Ramírez, 2014). In the 21st century, educational researchers need to determine what knowledge can be generated from education and its usefulness, to turn it into that instrument of cultural and economic development that a large number of countries in the world seek (Green, 2010).

6.4 What Knowledge Does Education Generate?

It has not yet been specified what knowledge is needed to develop educational research and what knowledge can be generated from the field of education (Green, 2010). At the time of writing, there are innumerable philosophical controversies about what constitutes scientific evidence in education. Epistemologically, for something to be considered true knowledge, there must be some guarantee of its existence. There is a consensus that knowledge in the natural sciences, in the social sciences, and in the humanities is built on interactions between people. The fact is not whether knowledge is a property of the phenomena studied. The challenge is that once the new knowledge is generated, it is possible to communicate it to others so that it is known and understood. In education knowledge is much more debatable because of the absence of a common language (Smeyers, 2013). Several issues emerge in relation to the knowledge that can be generated from the field of education.

One possibility to understand and investigate education is to examine the usefulness of the knowledge needed in this field to make it effective, as in other professional disciplines. Educational practices and their policies are two components that affect the effectiveness of education (Green, 2010). At first glance, knowledge is to ask what knowledge educators need to be effective in their teaching. Education has to produce knowledge to train the citizens of a country. At university level, the function of university education is to produce social mobility of people, training for the world of employment, and preparation for life (Scott, 2014). To speak of education is to speak of the development of the human being. The knowledge that educators need to do their work is very particular to their educational settings, to their students, and to the reality of their schools (Berliner, 2002; Green, 2010). In other disciplines, the need is for general knowledge. The science that has to deal with the particularities of its contexts is more complex than the science that has to do with the common and general aspects of a system (Berliner, 2002). In education there is a need to distinguish various forms of knowledge for practice (Cumming, 2012):

Knowledge about educational practices: If educational research focuses on studying educational practices to produce learning, then what is the basis of these practices, what educational objectives are pursued, what procedures or protocols are needed to implement them effectively, with which student populations produce better results, what training educators need to apply with skill or what educational policies are necessary for effective implementation.

Knowledge that comes from professional experience: This knowledge represents the wisdom developed by educators through the practice of the profession. This knowledge can come in the form of opinions, values, attitudes, or behaviors. It does not take much time in teaching to begin to identify the differences between groups even though they are the same ages and in the same academic grades. It does not take much time either to understand that sometimes it is necessary to deviate from the daily plan of education to respond to the particularities of students as a group in relation to the subject of teaching. This knowledge has two possible uses in educational research: (a) it can enrich knowledge about educational practices and (b) it can serve to validate the knowledge generated about educational practices, according to the educational contexts from which they come.

Knowledge of the educator's educational specialty: This is the knowledge acquired in universities, which is taught in the classroom, which is embodied in the curricula and its relevance is determined in the light of the educational realities of education systems (Green, 2010).

If educational research approaches are defined in the light of the usefulness of knowledge needed to have effective educational practices, administrative practices, or educational policies, then this should yield more reliable knowledge about education.

A second look at the usefulness of knowledge needed in education comes then from understanding how educational actions impact the student (Cumming, 2012). Gil-Cantero & Reyero (2012) identify three levels of knowledge that can be generated from educational practices:

Changes in the biological physical dimension (level 1): Effects of education on the improvement of the student in the psychomotor field and their physical well-being. Empirical studies on this level contribute to the design of intervention, psychomotricity, or health programs.

Changes in the psychological and sociocultural dimension (level 2): Effects of education upon the emotional and cultural development of students.

Studies that would seek to measure the effect of teaching strategies on the learner or measure behaviors.

Changes in the anthropological, spiritual, transcendent, and sense dimension (level 3): Effects of education upon the development of individuals as a social group of a country. Studies that would seek to explain how education should be and not how it works. Empirical research has no scope in this educational level. This is the type of research that would allow the development of a humanizing education. Understand the ability and tact of teachers to address the particular needs of their students. This is the level where theories are needed and where there are many philosophies and a great diversity of positions among educators (Cumming, 2012).

6.5 Validating Knowledge in Today's Education

Educational research is largely ex post facto. Ex post facto research sometimes resorts to the lives and experiences of the protagonists of education to generate knowledge. This raises questions about the possibility of knowing reality and human behavior (Clark, 2011). As a science, there are authors who argue about the need for educational research to increase the accuracy of the data it generates (Labaree, 2004; Marley and Levin, 2011; Koichiro, 2013; Smeyers, 2013). To achieve this, the opinions of the participants must be eliminated from the studies to demonstrate the facts of education (Marley & Levin, 2011). No data or statistics should be interpreted outside of reason, logic, and the norms of education (Koichiro, 2013). Epistemologically, for something to be considered true knowledge there must be some guarantee of its existence. The three considerations for validating knowledge are the following: (a) there is a belief that it exists, (b) there is a justification that guarantees its existence, or (c) it exists (Smeyers, 2013). In education, reality is neither predetermined nor stable (Thompson, 2012). Educational systems can make every effort to standardize the curriculum, teaching, and assessment methods, or the physical environment of the campus, but the individuality of the student emerges in their interpretation of the education they receive, in their ability to learn and in its rhythm of maturation. Knowledge in the natural sciences, social sciences, and the humanities is built on interactions between people (Thompson, 2012). Therefore, educational research has the challenge of confirming the authenticity of the information it generates. Educational research needs to eliminate from its studies the opinions to demonstrate the facts if it aspires to be scientific. It needs to replicate further studies as a form of validation and confirmation of their knowledge (Marley & Levin, 2011).

The issue of generalization of knowledge is recurrent in the allocation of funds for educational research. The following are the following consensus on generalization in educational research: (a) it cannot be the search for laws of universal application; (b) does not imply that the knowledge of a particular situation can inform another similar situation, and (c) the deep description of the educational phenomenon under study is fundamental to be able to identify the similarities and differences that allow the knowledge of a situation can help to inform another (Schofield, 2007). In the field of education, students' educational practices, policies, and behaviors are studied. So, generalizing in this context entails the following considerations (Phillips, 2005):

1. Human behavior can be influenced not only by the biology of the person, but also by their culture. Education changes culture. So, the search for these cultural patterns is necessary to be investigated and discovered.
2. It is argued that educational research is contextual. There are three reasons that indicate it is not: (a) ethnographical studies behaviors in contexts but it is considered a science other than the positivist and generalizes; (b) generalization always implies some kind of inference when applying. There are numerous arguments about the investigation of individuals or individuals, but many of the phenomena in education occur and are investigated in groups. Generalization must be parallel between individuals and between groups as in the physical sciences; and (c) a considerable part of educational research aimed at understanding the phenomena under study need not be positivist, but postpositivist.
3. The generalization of the findings can increase if the following aspects of knowledge production and their precision are considered: (a) definition and careful delineation of the phenomenon under investigation; (b) determine the functions and structures of the relationships being studied; (c) identify factors that are causal in the situations being investigated; (d) distinguish which factors are causal and which are by chance; (e) accurate measurement; (f) develop and validate theories and hypotheses; and (g) connect causes with effects.
4. The recommendations in clause 3 can assist in the accumulation of knowledge. In education, the important findings of a decade may lose their full relevance in the next decade and with the next generations of students because socioeconomic, cultural, or political conditions change. In the physical world, many findings last for life. This facilitates

the accumulation of knowledge. In education, social, cultural, and political conditions can make data lose relevance (Berliner, 2002). Defining the educational context and the phenomenon should help to better understand the duration of the data.

6.6 Conclusions

The premises of science are two: (a) that reality is intangible. This does not mean that it is superficial. (b) Knowledge is not easy to achieve, but the human being must fight against superstition and simple solutions that the mind presents (Gil-Cantero & Reyero, 2014; Alonso & Gil-Cantero, 2019). In contemporary culture of scientific research to generate knowledge, procedures are essential to differentiate between scientific knowledge and ordinary or daily knowledge of daily life. The result is the belief that the method controls research actions. The two limitations that arise from generating objective knowledge are the following: (a) each conclusion is restricted to the method and the premises that generate it, and (b) each research is developed from the perspective that the researcher selects to approach the phenomena which they study and the interpretation it gives them (Thompson, 2012).

In the 21st century, the selection of the research method must be aligned and respond to the scope it presents to capture the complexity of education and generate the data that is needed. Educational research does not have to be limited to experimental research to be scientific as long as a solid case is developed or raised (Phillips, 2005). Human complexity is of such magnitude that educational research has to take place in broader contexts where the real product of education can be appreciated (Thompson, 2012). This has two implications:

1. *Research in educational institution*: The value of educational and managerial practices is that they have to occur in educational settings where it is possible to observe, describe, and measure the effect of these on education and institutional effectiveness. This research informs the creation of educational policies and the decision-making of all its constituents (Ravish, 2014; Ballock, 2019; Cloonan, 2019). There is a need, and it is possible to conduct empirical studies to prescribe what education should be and understand its nature. Such studies should occur as close to the classroom as possible. The desirability of an empirical research on education will depend on the particularities of

each educational system that generates this prescription according to educational contexts (Mejías, 2008).

2. *Research outside the context of educational institutions*: There is a need to develop educational research away from the classroom where the products of education are actually observed. Educational research must move away from educational contexts where education takes place to understand other relationships about how education occurs (Lee, 2010).

7

The Next Educational Technology

Jesús Valverde-Berrocoso and María Rosa Fernández-Sánchez

University of Extremadura, Spain
E-mail: jevabe@unex.es; mafernandezs@unex.es

7.1 Introduction

"Emerging" technologies and educational practices include digital devices and applications (hardware and software), as well as pedagogical approaches and methodologies. Therefore, they are constituted by tools, concepts, innovations, and advances. Differentiation between "technologies" and "practices" underlines the role of the social, political, economic, and cultural context in determining the "emerging" character. In fact, "emerging technologies and practices" can be specific for certain academic disciplines (e.g., apps for learning mathematics), have a prior tradition that makes them more manageable (e.g., open knowledge in the field of sciences) or possess educational possibilities that are more appropriate for certain purposes (e.g., Wikis and blogs for postgraduate training in higher education).

7.2 Characteristics of "Emerging" Technologies and Practices in Education

Veletsianos (2010) identified five features that define emerging technologies/practices in education. Firstly, what makes these technologies and practices "emerging" are the educational environments or "ecosystems" in which certain technologies or practices operate. The teaching-learning process is a sociocultural phenomenon that is embodied in certain contexts and influenced by specific educational approaches. In this respect, technology is socially conformed, as it incorporates a worldview, values, principles, and beliefs of the people who participate in its design and use. Students and teachers can accept or reject certain technologies or practices, according to whether they satisfy their needs and values. Therefore, the sociocultural factor makes them "emerging." For instance, the use of social networks is an emerging practice among teachers, which offers diverse usage possibilities according to the purposes it is to be used for (instructional, social, or organizational). A technology/practice maybe, at the same time and according to the context, "emergent" or "nonemergent." Thus, e.g., online university training is a fully established model in distance education institutions but begins to be an "emerging" practice in traditionally " face-to-face" educational organizations.

In addition, the concept "new" is not a synonym of "emerging" and all technologies that are not currently used in educational institutions could be considered "emerging." Emerging technologies and practices can arise from recent innovations (e.g., three-dimensional [3D] printing or cloud computing) or from others that are not so novel, e.g., the use of free software platforms for virtual teaching-learning environments, such as Moodle). Some technologies, such as "virtual worlds" were identified as "emerging" in the last decade of the 20th century, and are still considered as such. Consequently, "novelty" is not an attribute that identifies the degree of "emergence" of technology or educational practice with precision.

Additionally, "emerging" technologies and practices are characterized by being "evolutional," i.e., by being in a dynamic or continuous state of transformation. It can be seen, e.g., that both technology and practice developed in a social network, such as Twitter, are in constant change, such as, e.g., in information filtering mechanisms as well as interaction with other users, forms of use for educational purposes, where microblogging technology has emerged as a new practice for literacy.

Another common element of "emerging" technologies/practices is the lack of knowledge about its implications in the teaching-learning process of students, teachers, and educational organizations. Educational research offering evidence on the effects of these innovations in education is lacking. When a new technology appears, there is always some initial research that can either adopt optimistic (utopian) or pessimistic (dystopian) positions and is limited to describing characteristics and supposed educational possibilities, without offering rigorous evidence on educational impact and implications. Most of this research carries out case studies that show preliminary approximations to try to understand these phenomena. The absence of studies on these "emerging" technologies/practices leads to the first uses being those of replicating processes already previously applied. Such as the case, e.g., of "virtual classrooms" on e-learning platforms (e.g., Moodle), which are initially used by teachers as mere repositories of text documents, replacing the photocopier, while ignoring the communicative and collaborative possibilities they offer to the educational community.

Finally, "emerging" technologies/practices generate expectations of change, which are mostly unfulfilled. Although the potential for educational innovation is acknowledged, such as, e.g., improvement in communicative interaction or contribution to educational equality, a process to provide the expected results is not completed. Educational institutions assimilate changes very slowly for organizational, cultural, and historical reasons. Discourse favorable to educational innovation is easily accepted, however, the same unanimity is not achieved in the application of the practices to materialize those ideas and intentions.

7.3 New Theoretical Approaches to Educational Practice with Emerging Technologies

It is necessary to identify theoretical models to guide the incorporation of emerging technologies/practices in the new educational ecosystems. The Internet has created a context for teaching and learning that is radically different to educational environments that existed before its creation, although it still contains some "evolutionary genes" from prior cultures and technologies (Anderson, 2016). For this reason, educational theories should be updated to contribute to the guidance of the pedagogical practice adapted to the educational needs of the 21st century. The following is a short review of some theories that could contribute to comprehending didactic action in the new ecosystems that are emerging in the field of education.

7.3.1 Trialogical Learning

The learning process has traditionally been explained through two epistemological metaphors. The first underlines the individual character of learning (monologic) and, consequently, adopts a computational model of the mind, where we learn by "acquisition" of information, by a cognitive, personal process, that is manifested in the "transfer" of knowledge. The second metaphor focuses on the social character of learning (dialogic), where "interaction" between individuals and mediators, inseparable from the context where it takes place, leads to a process of "participation" in different cultural practices and shared activities, generating a "distributed" knowledge (Paavola, Lipponen & Hakkarainen, 2004).

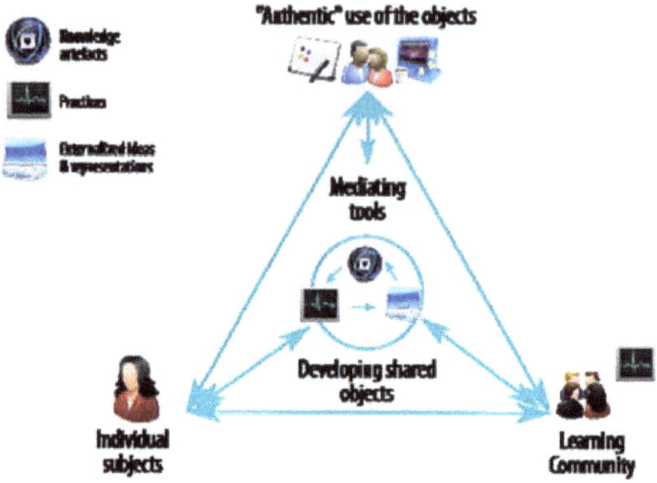

Figure 7.1 The concept of "trialogical learning."

Source: Prepared by authors, based on Hakkarainen & Paavola (2009).

Current, more complex, flexible, and connected learning ecosystems require a new metaphor based on the "collective creation of knowledge" employing "objects of shared activity." This new conceptualization integrates three fundamental elements for understanding learning: collaboration, creative processes, and emerging technologies (Moen, Mørch & Paavola, 2012). "Trialogical Learning" is the result of interaction between individuals or groups that create "shared objects" (material or conceptual), characterized by being knowledge bearers (epistemological objects) of an ambivalent kind, as on the one hand they are stable, i.e., they represent what is known at a given moment, and on the other, they are changeable and are consequently incomplete and open to subsequent development that generates new knowledge (Malins, Gray & Aggelos, 2015) (Figure 7.1).

Learning is an innovative inquiry process in which the objective is to progressively refine the artifacts of knowledge and develop long-term processes to extend the knowledge of the community and its competences. The processes of "collective creation of knowledge" are viable in emerging educational environments enabled by digital technologies for collaborative learning, e.g., computer supported collaborative learning (CSCL). These emerging technologies/practices provide collaborative spaces for the creation, sharing, and development of "trialogical" objects (Hakkarainen & Paavola, 2009).

7.3.2 Expansive Learning

The expansive learning theory (Engeström, 1987) holds that human beings and their collective groups, regardless of age, are creators of new culture. This theory studies the processes that an activity system, e.g., an educational institution, develops to solve its internal contradictions by the construction and application, by itself, of a new form of functioning (Engeström, 2007). Expansive learning is an evolution of the cultural-historical activity theory, which begins with the concept of "zone of proximal development" by Vygotsky, on which the so-called "third space" is defined, which is the hybrid and diverse learning environment, characterized by the multiplicity of its contexts (formal and nonformal), voices and narratives, characteristic of our current societies (Gutiérrez et al., 1999). Its development is continued with Leóntiev's theory of activity, on which Engeström defines the "activity triangle" (Figure 7.2). Moreover, it culminates with the formulation of "expansive learning," which considers that the bases of knowledge are the different "ecologies" that integrate individual, societal. and cultural artifacts, and which give rise to networks of activity systems by means of social collaboration, as well as learning environments differentiated by their historical-cultural dimension.

Learning is understood as a process that implies a sequence of cyclical actions from abstract to concrete, where finding contradictions is the reason for the change. Expansive learning is the foundation of an intervention denominated training "change laboratory" that analyses the

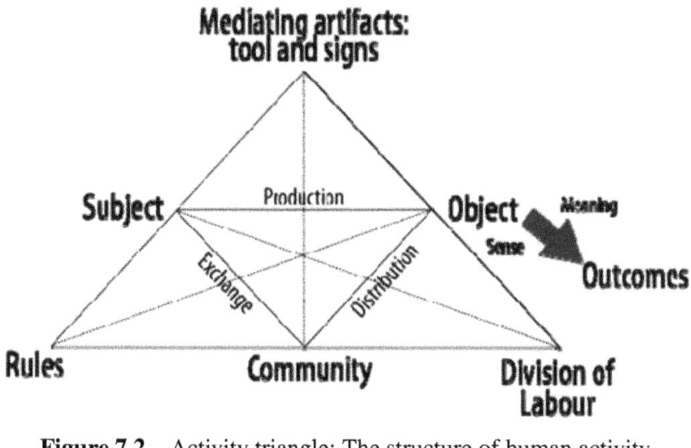

Figure 7.2 Activity triangle: The structure of human activity.

Source: Prepared by the authors, based on Engeström(1987, p. 78).

history, contradictions, and zone of proximal development of a group or organization, with the aim of explaining and guiding a collective transformation that will generate new forms of operation and a new culture, made with the support of emerging technologies with collaborative purposes (social networks, devices for communication).

The theory of expansive learning has given rise to a wide variety of didactic studies and innovations such as, e.g., the use of blogs in e-learning (Makino, 2007), the impact of educational reforms related to ICT in teacher training (Rasmussen & Ludvigsen, 2009), description of transforming models for using technologies in educational centers (Nleya, 2016) or the impact of the ICT on pedagogical discourse (Ingram, 2016).

7.3.3 Heutagogy

"Heutagogy" is the study of self-determined learning. It is a pupil-centered approach built on the foundations of constructivism and humanism. It takes learning as an active and proactive process, where pupils are the main agent of their learning, which take place as a result of their personal experiences. The teacher facilitates resources and gives guidance, negotiating with the pupil, "what" and "how" the pupil will learn. A key concept of heutagogy is "double-loop learning" (Figure 7.3) which takes place when students question and examine their values and beliefs as an axis of their meta-learning (Blaschke, 2012).

The underlying principles of heutagogy are the following (Blaschke & Hase, 2015, p. 81): (1) involve students in the design of their learning process in association with the teacher; (2) create a flexible curriculum so that it is possible to explore new interests and knowledge; (3) individualize learning as much as possible; (4) provide flexible or negotiated evaluation; (5) allow the student to contextualize ideas and knowledge; (6) provide multiple resources and allow the student to explore; (7) differentiate between acquisition of skills

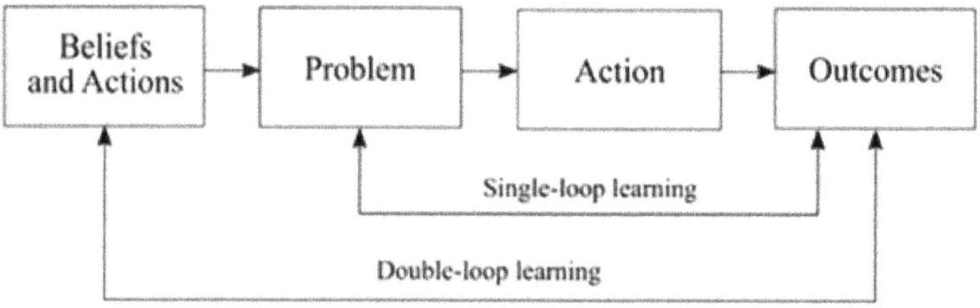

Figure 7.3 Double-loop learning (Eberle, 2009).

and deep learning; (8) acknowledge the importance of informal learning, which only needs to be enabled, not controlled; (9) trust the student; and (10) acknowledge that teaching can become an obstacle for learning. Therefore, it is necessary to facilitate more than to teach, to keep a distance and guide, providing a compass in preference to a map.

Emerging technologies/practices offer a series of possibilities that are coherent with the principles of heutagogy: pupil-centered learning, self-regulated, collaborative, and based on meta-learning. Among these technologies/practices are social networks, personal learning environments (PLE), skill-based curriculum, portfolios, gamification or flipped classroom (Blaschke & Hase, 2015).

7.4 New Pedagogical Models for Emerging Educational Practice with Digital Technology

The evolution of media and social networks has enabled the generation of new ecosystems for learning, which together with the abundance of information, have developed new forms of communication and knowledge exchange. This more horizontal, open, and participative environment is the origin of the "PLE" and "open education," two emerging paradigmatic and cohesive technologies/practices of the most relevant educational innovations and disruptions, such as, e.g., the DIYer movement (Do It Yourself), OCW (OpenCourseWare), xMOOCs (Massive Open Online Courses with a connectivist approach), or the OER (Open Educational Resources), among others.

7.4.1 Personal Learning Environments

In the context of a web that has evolved as a powerful tool and an immense source of information resources for self-learning, as well as the limitations of the e-learning platforms (LMS) to incorporate new possibilities of interaction and creation, arises the concept of "PLEs," which highlights the importance of informal learning in the development of social software (blogs, wikis, podcasting, social networks, RSS, microblogging, instant messaging, or virtual worlds, among others). A PLE is a representation of informal learning from a variety of resources and networks, available online as well as offline.

To contribute to the development of PLEs for students, the following levels of interactivity can be established (Dabbagh & Kitsantas, 2012): (1) Personal management of information: the teacher encourages the use of social media (e.g., blogs or wikis) to create a personal learning space by the creation of own content and its management for purposes of personal productivity or the organization of learning activities. (2) Social interaction and collaboration: the teacher promotes the use of social media for basic collaboration and sharing tasks (e.g., comments on entries in blogs or spaces for participation in a wiki), to facilitate the development of the processes of pupil's self-regulation and self-evaluation. (3) Aggregation and management of information: the teacher motivates the students to use social media to summarize and add information to the previous levels, with the aim of reflecting on their personal learning experience. For this purpose, e.g., they can view the history of a wiki or learn how to configure an RSS service in the personal blog.

7.4.2 Open Education

Open education is the result of a convergence of three factors related to emerging technologies/practices: (a) an infrastructure favorable for web-based learning; (b) a growing number of open educational resources (OER) available through this infrastructure; and (c) a culture of participation and knowledge sharing that facilitates personalized learning (Bonk, 2009).

Open education implies the encouragement of transparent, collaborative, and social learning experiences. It develops teaching activities such as the use of open software to enable access to all digital resources; the integration of free and open content and resources in the teaching-learning process; the promotion of copy left type licences in the production and dissemination of materials made by the students; knowledge of the implications of the copyright laws in educational activities; guidance of the personal learning networks toward collaboration and sustainable learning; creation of reflexive, motivating, and student-centered learning environments that incorporate a diversity of teaching and learning strategies; the modeling of attitudes of students regarding openness, transparency, connectivity, and responsibility in the use of the educational resources or the support of participation and the development of cultures for collaboration in education and society (Couros & Hildebrant, 2016).

7.5 Emerging Trends in Education

Emerging technologies/practices are not created in a vacuum, but are built from values, beliefs, and principles that are an inseparable part of themselves.

Some of the trends that experts consider of special relevance in the field of emerging technologies and practices are described in the following paragraphs.

The main issues dealt with in the First International Symposium on Emerging Technologies for Education (SETE) were the following: (a) learning based on playing; (b) personalized and adaptive learning; (c) language learning; (d) self-evaluation and peer evaluation; and (d) web-based learning; (Wu et al., 2017). As a result of the European project iTEC (Innovative Technologies for Engaging Classrooms), the emerging technologies/practices with the greatest promise for education were acknowledged (Cranmer & Lewin, 2017): learning based on playing and gamification, augmented reality, learning analytics, cloud computing, computational thinking, and neurodidactics.

The "horizon" reports, drawn up by The New Media Consortium since 2002, identify and describe the emerging technologies/practices with a potential impact on education. Digital technologies on their own cannot transform education, and therefore it is fundamental to identify emerging pedagogies on which to base future didactic innovations.

The principal emerging technologies identified for primary and secondary education, with expected application in the short- and medium term, are the following (Adams Becker et al., 2016):

(a) Makerspaces. Workshops offering tools and learning experiences are necessary to help people develop their ideas. They are developed in flexible educational spaces open to experimentation, problem-solving, creativity, and collaborative learning. They have diverse contents and purposes, but their common point is motivation to create things.
(b) Robotics. Design and application of robots or automated machines to carry out a collection of tasks controlled by computer programs. In the educational context, they are an emerging practice for the development of high-level skills such as divergent thought, strategic planning, and computational thought.
(c) Virtual reality. Digital environments that simulate the physical presence of people and objects, providing realistic sensory experiences. Among the most frequent uses of this technology are "virtual worlds" that generate replicas of real locations, where students and teachers, represented digitally in the form of "avatars," can develop social interaction and communication activities, acknowledgement of spaces, or experiments on virtual objects or phenomena.

(d) Analytical technologies. Any action carried out on the Internet enables digital systems to collect a large amount of user data, which can be analyzed for different purposes. “Data mining” in education has applications in the personalization of learning and pedagogical decision-making in different phases of the teaching-learning
process.

In the case of university education, the most important emerging technologies, which are expected to be adopted in the short- and medium term, are the following (Adams Becker et al., 2017):

(a) Adaptive technologies for learning. Consisting of online applications and platforms that adapt to the individual needs of the students while they learn.
(b) Mobile learning (m-learning). Including any learning process mediated by the use of devices such as smartphones or tablets. It is a modality that enables the development of the teaching-learning process anytime and anywhere.
(c) New generation E-learning platforms (LMS). It is necessary to be able to integrate new tools in the LMS to offer greater interoperability, the possibilities of personalization are increased, tools are provided for data analysis for improved evaluation of learning, options for collaborative learning are increased, and accessibility is improved, assuming the principles of universal design.
(d) The Internet of Things (IoT). Consists of objects that can communicate through processors and sensors and can transmit information collected on the Internet. In education, these emerging technologies/practices are being applied for the personalization (pace of learning, training evaluation, motivation for learning) and collaboration (interest groups, knowledge- sharing, peer evaluation).

7.6 Conclusions

Four changes are needed to address upcoming educational technology in education (Selwyn, 2010): (a) To overcome the “means-ends” way of thinking that implies questioning the idea that digital technologies have intrinsic qualities that grant them the capacity to generate certain “impacts” on education, if used in the “correct” manner. (b) Comprehend the “here and now” educational phenomenon; before speculating on the possibilities of technologies, to know "in situ" how they are used. (c) Examine the use of technologies from all perspectives of the varied contexts that embody and define educational technology (government, industry, educational system, family, etc.). And finally, (d) To develop comprehension and action, i.e., make educational technology "fairer" as well as more "effective."

8

From Information Literacy to Digital Competence

Ángel M. Delgado-Vázquez and Blanca López-Catalán

Pablo de Olavide University, Spain
E-mail: adelvaz@bib.upo.es; blopcat@upo.es

8.1 Introduction

Information management—understood as a global cycle which includes at least its location, access, organization, and exploitation—has always been one of the fundamental pillars in the teaching-learning processes, both in formal and nonformal environments. The breakthrough of new information and communication technologies has transferred to the digital media a large part of the tasks of production, storage, dissemination, localization, capturing, organization, and, finally, exploitation of the information in the generation of new knowledge.

This is why, from the perspective of the basic skills necessary for the management of information in physical and analogical environments, it is necessary to now add a series of capacities which make possible the correct poise of individuals in the knowledge society. Consequently, it is necessary to educate both those born before the digital revolution and those so-called digital natives for them to be able to actively participate and take advantage to the maximum of the opportunities which stem from this paradigmatic change, both in their role as citizens and in that of workers and, obviously, in that of students.

To address this work, it is recommendable to carry out an approach to what initially was called "information literacy" and its evolution toward the concept currently called digital competence, analyzing the components which, over the years, have been becoming especially relevant.

8.2 From Information Literacy to Digital Competence

Information literacy emerged as a concept in the 1970s, as Bawden (2001) points out, in a clear connection with educational reform, especially in the United States. In fact, its first mention and definition is attributed to Paul Zurkowski (1974):

> *People trained in the application of information resources to their work can be called information literates. They have learned techniques and skills for utilizing the wide range of information tools as well as primary sources in molding information solutions to their problems.*

From this first approach to "information literacy," numerous researchers and organizations, beyond the library area as well, have developed definitions mainly in the heart of initiatives, investigations, plans, programs, and frameworks.

Many authors have situated the basis of the concept in the classic programs of library literacy and bibliographic instruction developed by libraries for their users, which sought to train in the correct use of their collections (Arp, 1990; Rader, 1990). These actions transcended the mere localization of the documents, also enabling the students in the use, recognition of the nature of the information of each source, and in the handling of indexes and other description and synthesis instruments.

Various authors and organizations, especially in the area of information, have offered their own definition of information literacy. One of the earliest and most classic ones is that of the American Library Association (1989), "to be information literate, a person must be able to recognize when information is needed and have the ability to locate, evaluate, and use effectively the needed information"; which will be followed by others such as the Council of Australian University Librarians (CAUL), the Australian and New Zealand Institute for Information Literacy (ANZIIL), the Society of College, National and University Libraries (SCONUL), and the International Federation of Library Associations and Institutions (IFLA). Recently, the Chartered Institute of Library and Information Professionals (CILIP) have updated theirs.

All these organizations back including information literacy within the educational curriculum, with special emphasis on higher education. However, the need for information literacy is also recognized outside the library area. An example of this is that supranational organisms such as the UNESCO, the

Table 8.1 Outstanding models for information literacy.

Year	Models	Developers	Link
1990	Big6 Skills	Eisenberg and Berkowitz	https://thebig6.org/
1997	Seven Faces of Information Literacy	Christine Bruce	www.christinebruce.com
1999	The Seven Pillars of Information Literacy	Society of College, National and University Libraries (SCONUL)	https://www.sconul.ac.uk/page/seven-pillars-of-information-literacy
2000	Framework for Information Literacy for Higher Education	American Library Association/Association of College and Research Libraries (ALA/CRL)	http://www.ala.org/acrl/sites/ala.org.acrl/files/content/issues/infolit/Framework_ILHE.pdf
2006	Six Frames for Information Literacy Education	Christine Bruce	https://doi.org/10.11120/ital.2006.05010002
2012	CILIP information literacy model	Library and Information Association (CILIP)	https://www.cilip.org.uk

Source: Own elaboration.

OECD, and the European Union (EU) have developed their own initiatives in pursuit of the literacy of the citizenry.

Having recognized the importance of information literacy, models were developed oriented at information literacy teaching/learning, both general and adapted to different publics and/ or educational periods Some of the most outstanding (Table 1) are the Big6 Skills (Eisenberg & Berkowitz, 1990), The Seven Pillars of Information Literacy (SCONUL, 1999), Information Literacy Competency Standards for Higher Education (ALA/ACRL, 2000), the CILIP information literacy model (2012), and Seven Faces of Information Literacy and Six Frames for Information Literacy Education, both by Christine Bruce (1997, 2006).

Information literacy extends to any media in which information can be represented and/or stored. However, the intensive use of information technologies at the global level has attained all aspects of people and, therefore, a large part of the information is found in electronic forms and is accessible through the Internet. Added to this is that the quantity of data and associated information is growing exponentially, creating problems of control, but also of access, even overloading (infoxication, information overload). These present new challenges which surpass the traditional information literacy when the previous models are still being developed.

The materialization of the term digital literacy is traditionally attributed to Gilster (1997) who defines it as:

> *The ability to understand and use information in multiple formats from a wide range of sources when it is presented via computers. The concept of literacy goes beyond simply being able to read; it has always meant the ability to read with meaning, and to understand. It is the fundamental act of cognition. Digital literacy likewise extends the boundaries of definition. It is cognition of what you see on the computer screen when you use the networked medium. It places demands upon you that were always present, though less visible, in the analog media of newspaper and TV. At the same time, it conjures up a new set of challenges that require you to approach networked computers without preconceptions. Not only must you acquire the skill of finding things, you must also acquire the ability to use these things in your life, the ability to understand and use information in multiple formats from a wide variety of sources when it is presented via computers.*

From this definition it follows that this is not a question of a new literacy but, rather, of a reformulation of information literacy to apply it to the digital medium (exclusively) and logically supporting oneself through the knowledge of the correct use both of the technological tools and platforms, and the networks and media which serve to produce, disseminate, and access information. Gilster points out a series of central components and competences in digital literacy which is worked on by Bawden (2001). Here he identifies the need to overcome the scheme of information literacy in the search for one that is greater and which includes not only a correct handling of the information (as a global vision) in an electronic format (identification, search, localization, access, organization, and reuse), but also everything related with the use of the technologies themselves (computer or ICT literacy); of the networks which serve to access information (Internet literacy); of the use and the comprehension of the new media and forms of communication, as well as the capacity to analyze messages and their context (media literacy). As can be seen, digital literacy is conceived as a set of other literacies and sets of skills.

For Bawden (2007) it is a question of joining together at least four components as listed in Figure 8.1.

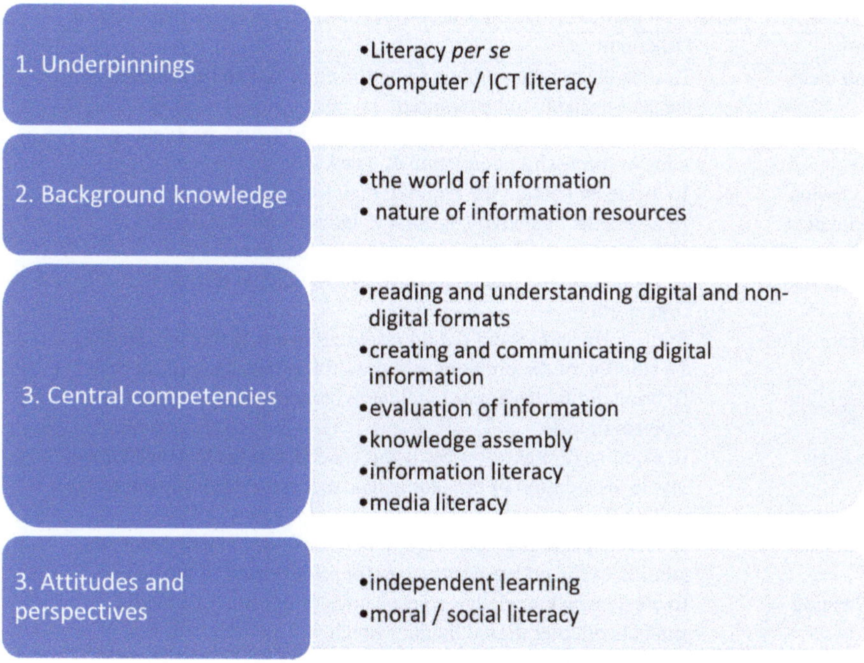

Figure 8.1 Components of digital literacy according to Bawden (2008).

Source: Modified from Bawden (2008).

To the scheme of components proposed by Bawden, other aspects are also added that are not strictly skill based and that, in his opinion, must previously possess digital literacy (the case of the underpinnings). Something similar can be said about the knowledge of the world of information and the nature of information resources. The third component has already deployed the competences and skills which, totally or partially, flesh out digital literacy. Lastly, the attitudes and perspectives are the evidence that the ultimate aim of digital literacy is to habilitate each person to learn by themselves what is necessary according to their personal situation (in direct connection with the idea of lifelong learning); and moral/social literacy refers to the need to understand the correct behavior in the digital environment, including questions of privacy and security.

Another of the widely diffused and accepted definitions of digital literacy is that of Martin & Grudziecki for whom it is a matter of "the awareness, attitude and ability of individuals to appropriately use digital tools and facilities to identify, access, manage, integrate, evaluate, analyze

Table 8.2 Processes for digital competence according to Martin & Grudziecki (2006).

Process	Descriptor
1. Statement	To state clearly the problem to be solved or task to be achieved and the actions likely to be required
2. Identification	To identify the digital resources required to solve a problem or achieve successful completion of a task
3. Accession	To locate and obtain the required digital resources
4. Evaluation	To assess the objectivity, accuracy, and reliability of digital resources and their relevance to the problem or task
5. Interpretation	To understand the meaning conveyed by a digital resource Organisation
6. Organisation	To organize and set out digital resources in a way that will enable the solution of the problem or successful achievement of the task
7. Integration	To bring digital resources together in combinations relevant to the problem or task
8. Analysis	To examine digital resources using concepts and models which will enable the solution of the problem or successful achievement of the task
9. Synthesis	To recombine digital resources in new ways which will enable the solution of the problem or successful achievement of the task
10. Creation	To create new knowledge objects, units of information, media products or other digital outputs which will contribute to task achievement or problem solution
11. Communication	To interact with relevant others whilst dealing with the problem or task
12. Dissemination	To present the solutions or outputs to relevant others
13. Reflection	To consider the success of the problem-solving or task- achievement process, and to reflect upon one's own development as a digitally literate person

Source: Martin & Grudziecki (2006).

and synthesize digital resources, construct new knowledge, create media expressions, and communicate with others, in the context of specific life situations, in order to enable constructive social action; and to reflect upon this process" (Martin & Grudziecki, 2006).

It is a question of a definition which conceives digital literacy in a three-phase process, the first of them being digital competence (Table 2). Centering on the first phase—digital competence—the authors establish an achievement of 13 sequential processes which would enable solving problems or tasks via the use of digital media.

The other two phases of digital literacy would be, on the one hand, digital usage: the application of the digital competence in contexts, domains, disciplines, or specific situations; and on the other hand, digital transformation, "when the digital usages which have been developed enable innovation and creativity and stimulate significant change within the professional or knowledge domain" (Martin & Grudziecki, 2006).

For its part, in the heart of the EU and in the context of the definition of the key competences for lifelong learning, digital competence is defined as "the confident and critical use of Information Society Technologies for work, leisure, and communication."

(European Parliament and the Council, 2006). This definition has been recently updated by the Council of the European Union's (2018) adoption of the Proposal for a Council Recommendation on Key Competences for Lifelong Learning, to adapt it to the DIGCOMP, the framework for the development of digital competence of all citizens in the EU whose frameworks we will analyze again.

Thus, the new definition is as follows:

> *Digital competence involves the confident, critical and responsible use of, and engagement with, digital technologies for learning, at work, and for participation in society. It includes information and data literacy, communication and collaboration, media literacy, digital content creation (including programming), safety (including digital well-being and competences related to cybersecurity), intellectual property related questions, problem solving and critical thinking* (Council of the European Union, 2018).

This new definition also again goes thoroughly into the meaning of digital competence itself including essential knowledge, skills, and attitudes related to this competence.

To sum up, the frenetic evolution of digital technologies, habits, and the penetration of technology in individuals has given rise to an evolution of the needs that we put forward at the beginning. Digital competence is conceived as an indispensable tool to support creativity and innovation. This competence must facilitate social inclusion, collaboration with others, and creativity toward personal, social, or commercial goals. Every citizen has to be aware of his/her opportunities and limitations and has to be capable of managing and protecting information, content, data, and digital identities, as well as recognizing and effectively engaging with software, devices, artificial intelligence, or robots. The following question is: how to empower people to satisfactorily adapt them to the new environment?

8.3 Frameworks and Programs for the Teaching of Digital Competence

The changes described until now, including those related with the definitions and the very nature of digital competence, have been reflected in the specification of different initiatives for acquiring the skills and abilities related with it. The same as in the case of digital competence, these programs have been undergoing transformations over time, passing, in some cases, from centering on information and/or IT literacies to evolving toward complete dedication to digital competence.

As an example, we present this evolution in three main supranational organisms: the UNESCO, the OECD, and the EU.

8.3.1 Framework of Digital Competence in the UNESCO

In 1996 a commission of experts headed by Jacques Delors delivered to the UNESCO a report that, with the title *Learning: the treasure within* (Delors and the International Commission on Education for the Twenty-first Century, 1996), reviews the current state of education in the world and proposes lines of advance and cohesion for the new circle which is about to begin. The report stands out by expressly mentioning that "*as the 'information society' is increasing the opportunities for access to data and facts, education should enable everyone to gather information and to select, arrange, manage and use it*".

Later, the United Nations commissioned the UNESCO for the development of programs linked with the Decade of Literacy of the United Nations (2003–2012). Having received the order, the IFAP (Information for ALL) program was started. This, in collaboration with the (*International Federation of Library Associations* and Institutions) and the ICA (International Council of Archives) and as many as 26 countries marked as one of its objectives "*to support training, permanent education and Lifelong Learning in matters of communication, information and IT*" (Wilson, 2019).

"Informational literacy" stands out among the priorities of the program. A definition is made of this in the project's website which raises it to the status of "a fundamental human right which fosters social inclusion in all countries" (Wilson, 2019). Among the activities most pointed out is the holding of various workshops which, under the title *Training-the-trainers in Information*

Literacy (TTT) (Boekhorst & Horton, 2009) means to train the trainers in information competences all over the world.

In Bangkok in 2010 the UNESCO began to lay the foundations for the development of a new program of Media and Information Literacy (MIL) (Moeller, Joseph, Lau & Carbo, 2011). One of the major novelties, and that the title of the program itself implies, is the inclusion along with information competences, of those related with media literacy, which have been talked about in previous documents and meetings.

The concern of the United Nations for literacy at a global level has again been spotlighted with the definition of the sustainable development goals, from the 2030 Agenda, where the point of the development of Target 4.4 *by 2030, substantially increase the number of youth and adults who have relevant skills, including technical and vocational skills, for employment, decent jobs and entrepreneurship*, is completely related with digital competence.

8.3.2 Framework of Digital Competence in the OECD

The first initiative is DeSeCo, (Definition and Selection of Competencies). This is a project begun in 1999 to select and define the essential competences for the life of people and the good functioning of society in which "the ability to use knowledge and information interactively" is included as a competence (C1).

Later, the OECD, through the Centre for Educational Research and Innovation (CERI), organized a congress centered on "the skills and competences of the 21st century for the learner of the new millennium." Its report established a framework of skills and competences for the 21st century divided into three dimensions: information, communication, and ethical and social impact. Within the dimension of information, the report makes two new divisions: "Information as a source: seeking, selection, evaluation, and organization of information" and "information as a product: the restructuring and modeling of information and the development of own ideas (knowledge)."

Currently, the OECD is redefining its vision of the principles which have to sustain the educational systems of the future through Learning Framework 2030. In its last position paper (OECD, 2018), it can be deduced that digital competence is going to play a central role in the definition of the OECD's strategy in the coming years:

"Two factors, in particular, help learners enable agency. The first is a personalized learning environment (...). The second is building a solid foundation: literacy and numeracy remain crucial. In the era of digital transformation and with the advent of big data, digital literacy, and data literacy are becoming increasingly essential, as are physical health and mental well-being."

8.3.3 Framework of Digital Competence in the EU

In 2006 The EU, through the parliament and the Commission makes a recommendation to the member states about the key competences for permanent learning (European Parliament and the Council, 2006) which includes a new referential framework of competences titled *Key competences for permanent learning: A European reference framework.*

This new framework again establishes a series of key competences for the personal and professional development of European citizens. Some of these competences spell out the knowledge, capacities, and attitudes which information literacy deals with.

Strategy Europa 2020 is started up in 2010. This is an initiative to improve the capacities of Europe within the context of the economic crisis and for it to emerge strengthened. Among the actions to be carried out it is decided to create a Digital Agenda for Europe (European Commission, 2010). This agenda came into being in the following months as an inclusive action plan to achieve "sustainable economic and social benefits from a digital single market based on fast and ultrafast internet and interoperable applications." Among the objectives which the agenda marks out is "*to foster digital literacy, training and inclusion*," and within this a series of key actions aimed at achieving it. As will be seen further on, information literacy is being included along with computer literacy within digital literacy.

In December 2010 the Joint Research Centre, on behalf of the Directorate General for Education and Culture, begins working on developing a framework for acquiring digital competence to respond to the challenges posed by the digital society. Fruits of these works are the two versions of the framework which have been published until now: DIGCOMP 1 and 2.0. DigComp 2.0 identifies the key components of digital competence in five areas which can be summarized as:

"(1) Information and data literacy: To articulate information needs, to locate and retrieve digital data, information and content. To judge the relevance of the source and its content. To store, manage, and organize digital data, information, and content.

(2) Communication and collaboration: To interact, communicate, and collaborate through digital technologies while being aware of cultural and generational diversity. To participate in society through public and private digital services and participatory citizenship. To manage one's digital identity and reputation.

(3) Digital content creation: To create and edit digital content. To improve and integrate information and content into an existing body of knowledge while understanding how copyright and licenses are to be applied. To know how to give understandable instructions for a computer system.

(4) Safety: To protect devices, content, personal data, and privacy in digital environments. To protect physical and psychological health, and to be aware of digital technologies for social well-being and social inclusion. To be aware of the environmental impact of digital technologies and their use.

(5) Problem solving: To identify needs and problems, and to resolve conceptual problems and problem situations in digital environments. To use digital tools to innovate processes and products. To keep up-to-date with the digital evolution" (Vuorikari, Punie, Carretero, & Van Den Brande, 2016)."

These are three of the initiatives which have a greater impact and/or projection for being managed by supranational organisms with a long trajectory. On the other hand, many actions are currently being developed in this sense at the national level from the report that the UNESCO Institute for Statistics (2018) has elaborated to contribute indicators which serve to measure Indicator 4.4.2 of the Sustainable Development Goals (2030 Agenda). This goes as far as to list 47, as well as gathering and analyzing the list, which, for its part, has already been the object of analysis in the elaboration of the DIGCOMP, the European framework (Ferrari, 2013).

9

VLE Environments and MOOC Courses

José Gómez Galán[1], Cristina Lázaro Pérez[2], Jose Ángel Martínez López[2], and Eloy López Meneses[3]

[1]University of Extremadura, Spain, and Ana G. Méndez University, Puerto Rico-USA
[2]University of Murcia, Spain
[3]Pablo de Olavide University, Spain
E-mail: jgomez@unex.es; jogomez@uagm.edu; cristina.lazaro2@um.es; jaml@um.es; elopmen@upo.es

9.1 Introduction

There are many forms of virtual platforms for e-learning (online distance learning) and b-learning (blended learning: classroom and online distance learning), known as *learning management system* (LSM) environments, which allow the creation of *virtual learning environment* (VLE). Through these, it is possible to recreate in the digital world not only all the possibilities and structures of the traditional classrooms and spaces of the university, but also to offer all the innumerable advantages of a new virtual dimension in which the ties marked by time and physical space do not exist.

Of all the web applications for VLE development worldwide, the one with the greatest extension and acceptance is Moodle (Bogdanovic, Barac, Jovanic, Popovic and Radenkovic, 2014; Gunasinghe, Hamid, Khatibi, & Azam; 2018; Alves, Miranda, & Morais, 2019; Baldwin, & Ching, 2020), especially in its open source software (OSS) version, but there are many other existing proposals, such as BlackBoard, Claroline, NetCampus, Phoenix Pathlore, Profe, Saba, SympoSium, Toolbook, Ucompass, VCampus, Virtual Training, Virtual-U, Web Course in a Box, WebBoard, WebCT, Whiteboard, etc. (Romero & Troyano, 2010; Gómez Galán, 2017d; Gómez Contreras, 2019). Although their structure and philosophy is very different, they all start from the fact of recreating a virtual environment that allows to carry out, with the maximum quality, distance learning processes. This is extremely useful to the current needs of universities, not only to carry out nonattendance training processes but also as a support to traditional teaching, offering new tools that were unthinkable until recently.

The development of ICTs in education has meant that practically all countries in the world, for one reason or another, have had to accelerate the process of incorporating and implementing them in schools, so that teachers have had to adapt, or even begin training, to use them in teaching practice. As Gallego and Alonso (1999) have already pointed out, the use of ICTs inside or outside the classroom does not mean that teachers are competent in their use or production. Training in them is essential.

We start from the fact that one of the great advantages offered by ICT is the amount of information that can be provided virtually, available to both students and teachers, quickly and easily. As presented by Ramírez (2013), this information with free access increased in such a way that it gave rise to multiple resource deposits, as is the case with *Open Educational Resources* (OER).

The OERs had their beginnings in 2001, when the Massachusetts Institute of Technology (MIT) created the *Open Course Ware* (OCW) program, and from that moment the interest in them grew. According to Santos, Ferran & Abadal (2012), open content for education has had two stages in its development: the first was focused on providing access to content, while the second was more concerned with its incorporation into educational practices.

In the last decade, all these digital spaces have been strongly developed and implemented in a large number of educational centers and, above all, universities at an international level, giving rise to various online learning environments and the creation of platforms or virtual campuses.

9.2 Moodle as a VLE Example

The most widespread today is the Moodle platform, which is freely distributed and offers an advanced system of virtual management of teaching-learning processes. Such platforms serve as a complement to both classroom-based classes to deposit topics or share ideas through forums, as well as to carry out exclusively online or blended learning training (Area & Adell, 2009; Dziuban, et al., 2018; Stein & Graham, 2020).

What is important, however, is not the information that is deposited on these platforms, but the use that is made of it by both teachers and students.

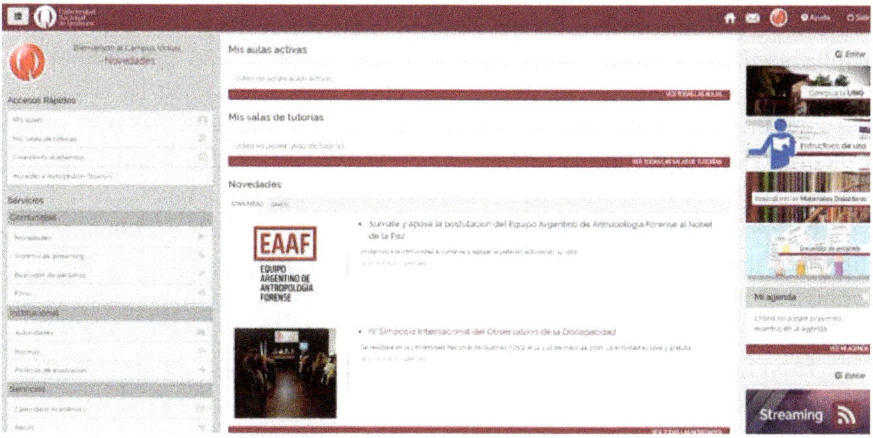

Figure 9.1 Ejemplo de campus virtual.

Source: https://es.m.wikipedia.org/wiki/Archivo:Vista_campus_nuevo.png

Authors such as Mirete, Cabello, Martínez Segura & García Sánchez (2013), argue that the use given to these tools does not depend so much on the type of support used but rather on what society demands at the time and the needs of the students.

Virtual campuses have generated a great impact not only at the educational level but also in the business environment, giving the opportunity to train any person anywhere, continuously and throughout life. All this is possible thanks to the information provided by the Internet and to each of the resources found on the network and on these platforms or virtual campuses, facilitating the relationship with the real world around us and the possibility of sharing knowledge. In addition, today they participate in multiple possibilities, from classic distance learning designs to complex immersive virtual reality systems (López Meneses & Gómez Galán, 2010; Wright & Reeves, 2019; Radianti, Majchrzak, Fromm & Wohlgenannt, 2020).

These virtual environments should be flexible and open, and should focus on the needs of the student body (Figure 9.1). Proper information and knowledge planning must be carried out, for which it is essential to use an innovative methodology (Salinas, 2012; Cassidy, 2016; Hamutoglu et al., 2020). These systems must also have adequate support to carry out a correct development of the platform and its maintenance. This requires good teacher training, as well as guaranteeing the quality of the educational material provided.

It was in 2012 when Moodle appeared in the technological field, providing innovative combinations of file management and educational resources such as forums or wikis. On the other hand, it also had its critics due to certain limitations such as the absence of creativity on the part of the teaching staff since, in many cases, its use was very limited and could not exceed or go beyond what the system itself provided (Delgado, 2013). However, since then its growth and development has been exponential, multiplying its functions and possibilities. Today it constitutes a high percentage of the world's virtual campuses (Kerimbayev, Nurym, Akramova, & Abdykarimova, 2020; Zabolotniaia, Cheng, Dorozhkin, & Lyzhin, 2020).

The Moodle platform has many features, and its main advantage is that it is free and an open source software (Figure 9.2). This allows teachers to fully customize the system to give it the form and content to best achieve their goals in line with the characteristics of their students. The basic functions of Moodle, which can serve as an example of other similar virtual platforms (Sobenis, & Torres (2019), are: (a) placing educational

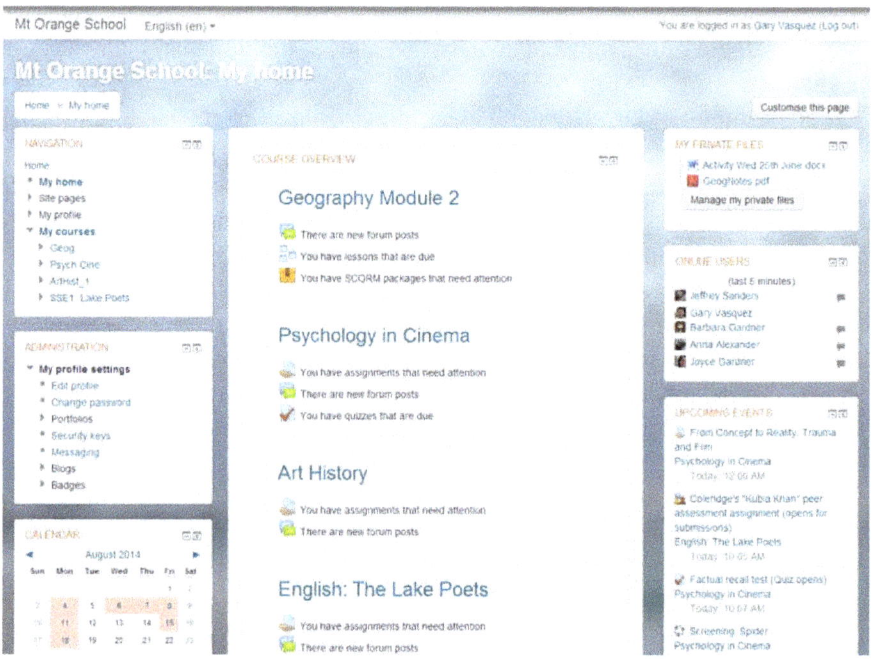

Figure 9.2 Training structure using Moodle.

Source: http://school.demo.moodle.net/

resources on the Internet to make them available and share them; (b) creating communication spaces such as forums, messages, and chats; (c) creating or placing automated interoperable assessments; (d) collecting, reviewing, and grading assignments; and (e) recording grades and other actions of participants. In addition, it is based on the following foundations: (a) students can share course and support materials from any location and at any time; (b) the use of communication tools, such as e-mail or instant messaging, can be used either in class or online; (c) it encourages collaboration between students and learning; (d) teachers can set assignments and assessments; and (e) the platform becomes an effective tool for addressing diversity of learning and developing intellectual competencies.

9.3 The MOOC Courses

Among the dynamics of change brought about by VLEs, the MOOC phenomenon is one of them. It is in this context of virtualization and digitalization of distance learning processes, and specifically in the field of higher education, that the MOOC courses (massive open online courses) were born. In line with Kregor, Padgett & Brown (2013), Siemens (2013), Gómez Galán, (2014a), Méndez, López, and Barra (2019), or Gómez Galán, Martín, Bernal & López Meneses (2019), this is an innovative training offer that is characterized by being (a) massive courses, in which a very high number of students from multiple countries and cultures can participate, in principle without restrictions of any kind, and that can even be scalable; (b) that allow for the creation of subnetworks depending on geographical location, prior knowledge, language, etc.; (c) they are usually free of charge, or with a minimum tuition fee; (d) they are available online and all activities are carried out in a virtual context; and (e) they have all the characteristics of structured courses, sequenced in an orderly fashion with a beginning and an end.

In a first instance, therefore, MOOC courses are interesting training modes to be developed in this polychromatic plurality of formal and informal educational contexts. This implies an innovative model of mass education that exploits in a paradigmatic way the potential and relevance that information and communication technologies have at present in society (Pérez-Parras and Gómez-Galán, 2015). A MOOC is a way to learn; ideally it is an open, participatory, distributed course, and a lifelong learning network; it is a way of connection and collaboration; it is a shared work (Vizoso-Martín, 2013); and with the intention of promoting virtual and personal learning environments (Barana et al., 2016; Vázquez-Cano, López-Meneses & Martín-Padilla, 2018).

Today there are many and varied types of MOOC courses (Figure 9.3). The main one is based on its *connectivist* characteristics, and a division can

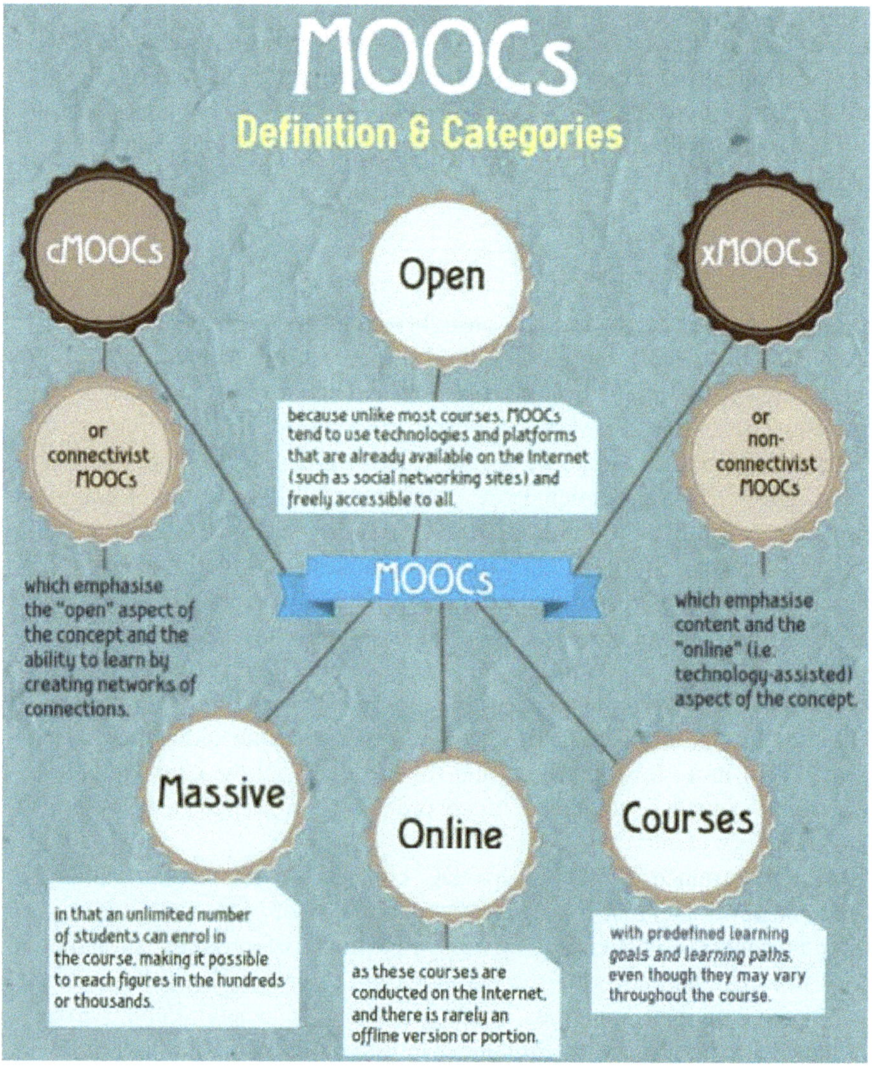

Figure 9.3 Definition and main categories of MOOC courses according to their relationship with connectivism.

Source: Universitat Oberta de Catalunya. http://openaccess.uoc.edu (customized image).

be established between cMOOC courses, based on learning in networks from the connectivist postulates, and xMOOC courses, which, although developed in virtual contexts, are based on traditional learning models.

A third type can also be given according to this classification: tMOOC courses. This would be a hybrid model of the previous ones, whose fundamental element is the performance of tasks by the student (Pomerol, Epelboin & Thoury, 2015; Fidalgo, Sein & García-Peñalvo, 2016; Rincón, Mena, Ramírez & Ramírez, 2020). However, in practice, there are not many differences in results or research that can demonstrate that these typologies are actually applied based on their theoretical assumptions.

Connectivism, so closely linked to MOOC courses, is understood as an apparent theory of learning in which, above and beyond the content, what is relevant are the flows of information that are generated and diversified in a context of people interacting (Siemens, 2005; Siemens 2013). However, and although it has many followers, there are studies that question the supposed benefits of connectivism. Bell (2011) or Clara & Barberá (2013), e.g., consider it very problematic from an educational perspective, and advocate a more appropriate methodology with fewer drawbacks. This is why, at present, MOOC courses are still evolving and other typologies and classifications are being proposed according to their characteristics and objectives.

9.4 Advantages and Disadvantages of MOOC Courses

This new modality of knowledge expansion can help transform classrooms, which are limited in time and sometimes reserved for a social elite, into new ubiquitous, connective, informal, and horizontal learning scenarios that can facilitate the digital inclusion of the most disadvantaged and the birth of interactive virtual communities of collective intelligence.

Similarly, the MOOC can be considered by universities as a strategic investment, since they are considered as indicators of technological innovation for learning, drivers of brand value, means of dissemination of the educational offer of universities, projection of the signs of identity of each university, and attracting students and fulfilling their social purposes (Méndez-García, 2013; Vázquez-Cano & López-Meneses, 2014; Salas, Morales, Villota & López Meneses, 2019).

The new didactic paths with the objective of learning to learn in a massive way can constitute a new techno-social trend, especially oriented in the panorama of higher education to energize university innovation, or simply derive toward a new business model for universities and institutions without demonstrated quality (Zapata, 2013; Vázquez-Cano, López-Meneses & Sarasola, 2013).

MOOCs, in general, are excellent tools for disseminating research and knowledge generated in the universities themselves, establishing learning communities within them and even identifying people and institutions for future collaboration. These training trends could also be used to generate both initial and continuous teacher training processes. The massiveness of this type of training may mark a before and after in the coverage of teacher needs, especially in Africa and Asia, where it is most needed (Peña, 2014).

The main drawbacks of the MOOC phenomenon, on the other hand, arise from an essential questioning of the philosophy with which they were born: free and massive. These two adjectives characterize and give specificity to this type of training, but their materialization makes it difficult in many cases to combine them effectively according to the pedagogical and economic approach from which they start.

Schulmeister (2012) states that the critical points of the MOOC can be summarized as high dropout rates; lack of feedback and low interaction; no reliable verification of learning outcomes and assessments; and that a great diversity of topics predominate, but without an explicit curriculum. Also, voices have been presented that speak of a process of *mcdonaldization* of education through the distribution of standardized educational packages worldwide (Aguaded, Vázquez-Cano & Sevillano, 2013; Blas, Morales, & López-Belmonte, 2019). In this sense, it is indicated that the MOOCs start from a pedagogical design that moves them far away from the didactic principle of connectivism and group work, and can be labeled as *impoverished e-learning*. Similarly, as they are free and focused from the beginning of massiveness, contact is lost with the tutor, who becomes a moderator or forum reviewer in the best of cases. The MOOC course would thus become a series of short videos and/or documents linked under the thread of a theoretical index in which the student through their viewing and small self-assessments learns almost autonomously.

In most of these courses, therefore, there would not be adequate processes of tutoring, content discrimination, student intervention, and teamwork. It is also very important, and in what must be improved, the evaluation of the intrinsic quality of these courses, for which outstanding advances are being made with instruments and observatories for their evaluation (Baldomero, 2015; Baldomero and Leiva, 2016; Gómez Galán, 2020).

Nevertheless, and in general, using all their potential and overcoming the possible limitations they may have, MOOC courses are an excellent resource that allows them to reach everyone, and all ages, with high quality content generated in the context of higher education, and represent a milestone in the processes of virtualization and digitalization of education.

9.5 Structure of a MOOC Course

MOOC courses participate in the processes of knowledge elaboration in groups and individuals within the digital paradigm, the open nature of knowledge supports (open access) or learning resources (OER). They must be approached from this perspective and designed within these parameters. Ideally, they should be designed as follows (Vázquez Cano, López Meneses, & Sarasola, 2013):

Phase 1. Video presentation of the fundamental aspects, structure, objectives, and contents of the MOOC course.

Phase 2. Presentation of teachers and students in a discussion forum.

Phase 3. Approach and explanation by the MOOC course tutor of a practical case that requires a collaborative and participatory approach for its solution, which presupposes a solution that can be applied in real life in a social, academic, or professional area.

Phase 4. Learning can be made relevant if it is contextualized and participants are able to relate the concepts to their real-life application. The objective of the design of the case study is to present the participant with a real or simulated case so that they are able to solve it as they progress through the course contents. The case study will use a research-based approach to stimulate critical thinking skills and allow participants to interact with the content, thus creating an engaging and meaningful learning experience. In addition to acquiring relevant knowledge and skills, the holistic approach to learning, especially in the context of lifelong learning, should also include the exchange of knowledge and experience to enhance participants' previous experiences. For this purpose, the following educator-led activities are recommended:

- One or more discussion forums. The publication of one or more discussion questions after the first modules to facilitate constructive debate around the questions. The forum can also invite the participation of professionals relevant to the subject matter of the course who can answer the questions posed by the participants.

- Other challenges. All additional content in the form of videos, examples, etc., that the instructor and the participants of the course share with the group can be posted in this section. The trainer can use this opportunity to present new challenges to the participants (in the form of a collaborative group) based on the published material.

Phase 5. The course ends with a final session in which the teachers-instructors provide the students with all the feedback accumulated during the training action and invite them to continue reflecting on social networks and discussion forums developed for this purpose after the course.

Similarly, it is worth mentioning that the new trends in MOOC courses encourage users to be content generators, rather than passive participants, and to maintain horizontal communication with teachers (García, Tenorio & Ramírez, 2015). This requires that these new MOOC models harmonize the interests of individuals and institutions to create collaborative learning communities (Testaceni, 2016; Gómez Galán, Lázaro, Martínez López & López Meneses, 2020).

9.6 Main MOOC Course Platforms and Resources

Today, there are countless reforms in which MOOC courses are developed, since practically all higher education institutions offer them. However, it is possible to present a list of the main ones, with more tradition and experience, and it offers a very wide and permanently updated catalogue of courses. We could highlight the following (López Meneses, 2017; Gómez Galán, Martín, Bernal & López Meneses, 2019):

- Coursera (www. coursera.org). It is founded by two other Stanford professors: Andrew Ng and Daphne Koller. After the departure of Sebastian Thrun from Stanford to found Udacity, Coursera was born in 2012 as a company "committed to bringing the best education for free to anyone who seeks it." Today it counts with the participation of dozens of prestigious universities around the world (Figure 9.4). The courses, which have a start and end date, are launched periodically and contain video lessons that are published weekly. The lessons feature interactive exercises. An innovative aspect was that some courses offered peer review, which allowed other students to participate in the evaluation of the content presented by other students.
- edX (http://www.edx.org). The edx.org website brings together free courses from several of the most prestigious universities in the United

Figure 9.4 Coursera platform.

Source: coursera.org

States, including international ones (Figure 9.5). It was founded by the MIT and Harvard University and in 2012 was joined by Berkeley University. Once the courses have been completed and the exams passed, the certificates are awarded with the characteristic of the name of the corresponding university. In some cases, they offer to buy a textbook, but it is not mandatory, and it is sent via mail to the student. They offer courses for all levels: beginners, intermediate, and advanced, and most of them are written in English although they are already offered in other languages.

- Udacity (http://www.udacity.com). Udacity was born out of an experiment at Stanford University in which Sebastian Thrun and Peter Norvig offered a course on artificial intelligence (Figure 9.6). The course achieved more than 180,000 enrolments. It was the embryo of a platform that today brings together many international universities. The courses can be selected according to different categories and levels.
- MiríadaX (http://www.miriadax.net). MiríadaX is an online training project that was launched in 2013 by Banco Santander and the Telefónica company, through the Universia Network and the Telefónica network (Figure 9.7). This project aims to disseminate virtual training to Latin American universities, and is the international reference in MOOC courses in Spanish. Many universities in Latin America are participating in this project.

Figure 9.5 edX platform.

Source: edx.org

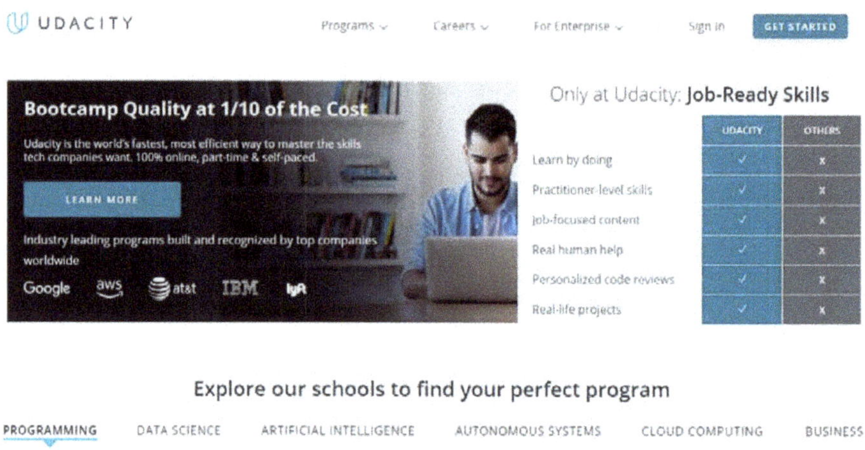

Figure 9.6 Udacity platform.

Source: udacity.com

On the other hand, it is worth mentioning other services that offer searches for the following:

MOOC courses:

- My Education Path (http://myeducationpath.com/courses). One of the aggregators that allows you to search for courses on some of the major MOOC platforms. My Education Path defines its mission as helping

Figure 9.7 MiríadaX platform. *Source*: miriadax.net

Figure 9.8 "My Education Path" search engine.
Source: myeducationpath.com

to find free alternatives to high cost university courses (Figure 9.8). In addition to this feature, My Education Path offers the ability to search for test centers that certify knowledge through MOOC courses.

– Class Central (http://www.class-central.com). On the home page, Class Central displays a space to perform a course search (Figure 9.9). There is also a list of the courses that are going to start next, where the name of the course, the name of the instructor, the area to which it belongs, the starting date, its duration, and the name of the platform that offers it are shown. Class Central currently displays courses from all major US MOOC platforms.
– No Excuse List: http://noexcuselist.com. Allows you to locate courses hosted on another group of platforms (Figure 9.10). To see the complete course directory just follow the link by clicking on the word "here." Many of the most popular educational platforms appear in this directory, organized according to the educational field to which they are dedicated.

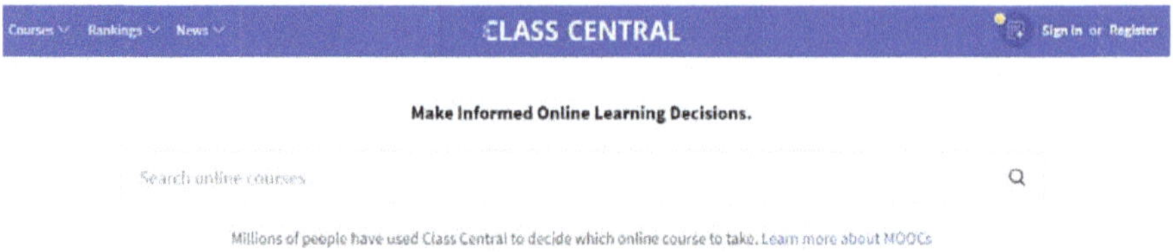

Figure 9.9 "Class Central" search engine.

Source: class-central.com

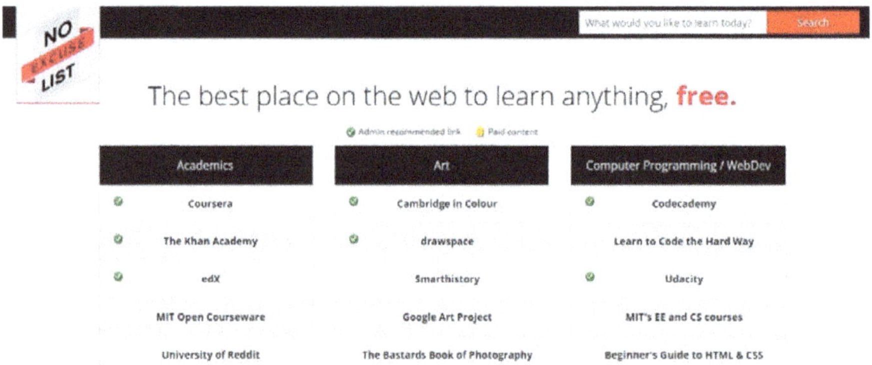

Figure 9.10 "No Excuse List" search engine.

Source: noexcuselist.com

9.7 Conclusions

VLEs are, without a doubt, an important part of the education of the future. Their possibilities are enormous, in a process of virtualization and digitalization that is unstoppable today. It is essential to train teachers in these new environments that are already a reality today. There is an urgent need to continue innovating in VLE, such as Moodle, which constitute the main framework of today's digital campuses. Only through new practical experiences that encourage interaction between these new digital phenomena will it be possible to make progress. And they must be generated and encouraged from the universities themselves. Today, unfortunately, there are still many limitations in this respect but impressive progress is being made.

Within this very broad panorama, we can highlight the MOOC courses, which are currently a reference point. They imply a reality of great dimensions and their presence in the formative and academic processes of the university may only have just begun. They can anticipate important advances in learning methodologies and evaluation processes, and even in educational management and policy, but for their true potential to be realized, a complete transformation of what we understand by higher education today will be necessary.

We are talking about new contexts that, in short, imply a dialogue between formal and nonformal education, on how to merge the most innovative models of e-learning and b-learning. And all this in a society, let's not forget, in which precisely the processes of informal education are increasingly important, thanks to the digital revolution, and which are conditioning its own nature. The challenge is not only to improve the quality of higher education but also to reach a much more important milestone in the history of education and culture: to make knowledge and science universal, within the reach of all human beings (Gómez Galán, 2020). And this requires new educational models which, precisely in the midst of the techno-media convergence, are within our reach.

10

Learning Communities

Rosabel Roig-Vila and Juan Francisco Alvárez-Herrero

University of Alicante, Spain
E-mail: rosabel.roig@ua.es; juanfran.alvarez@ua.es

10.1 Introduction

It is a constant feature in the educational world that, after some time has elapsed, those working in that area look for new (or not so new) pedagogical models able to meet the needs for change. An innovation or change is thus sought toward an improvement with respect to the previously established model, either because it comes to be regarded as obsolete or because people have grown tired of using it, or simply because it is simply considered that what used to work no longer does.

In the 1980s, an attempt was made to find a response to the need for a change in education in the United States. The model pursued had to enhance students' learning, to prioritize shared work—collaborative work—to a larger extent, and to succeed in improving students' academic results, as well as coexistence in classrooms. It was precisely some programs and initiatives with such an orientation which arose at the time that eventually resulted in what has come to be known as "learning communities" nowadays.

Quite a few years have elapsed since its irruption into the educational arena and, far from disappearing, the model referred to learning communities has taken a new and great impulse, both by the good results obtained wherever it is implemented and by the emergence of new variants. An example of the latter can be found in the so-called "virtual learning communities," which have paved the way for new ways of generating collective learning in new technological environments.

This chapter will help us to show the difficulty involved in defining and delimiting what a learning community actually is. After all, it cannot be fully understood unless we acquire a deep knowledge of each reality, context, group, or community that decides to follow such a model. Because of this initial complexity, its implementation is no easy task either.

10.2 The Concept of Learning Community

The concept of learning community has several interpretations. Even so, one can find a definition of this concept which adapts as much as possible and corresponds to a generalist vision of it. In the broadest meaning of this term, a learning community is a group of people with shared goals and attitudes aimed at improving the learning of all its members. Hence why it is also referred to as *community of learners* (Brown & Campione, 1994), since the objective sought by the individuals belonging to it consists in taking advantage of the experience distributed inside that community, so that sharing it can eventually improve their own learning.

Another important objective of learning communities is to foster the integration of learning and involving students in great questions which exceed the limits of the classroom (Kuh, 2008). On other occasions, learning communities set themselves as their priorities to overcome school failure and to improve coexistence, removing inequalities, and accepting everyone with dialogue and solidarity as referents to achieve it (Flecha & Puigvert, 2002).

We are consequently speaking about an educational model which seeks a social—and simultaneously educational—transformation that affects the whole educational community. This is a social and cultural change that will be accomplished with an integrated, participatory, and permanent education (Elboj, Valls & Fort 2000). It can additionally be established at any educational stage, whether it is primary, secondary, or university education.

Using the term "learning communities" with so many possible variations is making it lose its own meaning (DuFour, 2004); however, four key factors can ultimately help to define it (McMillan & Chavis, 1986):

- Membership or the feeling of belonging to a group, to which we express our loyalty, and where we work, share and help all its members, since everything that each member does has a bearing on the good of the group.
- Influence, because it implies an active—and not only reactive—participation.
- Compliance, since a learning community provides the chance to meet the individual needs of each one of its members.

- Shared events in which a conversation takes place and personal opinions are expressed, specific help or information is requested, and emotional links are built.

This transforming educational design model on which learning communities are based represents a stimulus in the search for an education meant to fight inequality by means of dialogue, cooperation, and solidarity; an education which seeks not only the involvement of every possible agent—including the direct agents (family, teachers, and students) and the indirect ones—but also the collaboration with official bodies, public and private entities, and anyone else interested in the actual community, be it a professional or a volunteer.

10.3 History

The first evidence recorded about the implementation of what could be called a "learning community" in an educational center dates back to 1968. The promoter of the School Development Program (SDP) was James Comer, from Yale University, who carried out this experience between 1968 and 1975 with good results at the public schools located in the New Haven region (Comer, 1980). The SDP was basically different from the pedagogical models of the time, both from the already existing ones and from those which arose at that moment, in the following aspects:

- it paid special attention to a mental health program focused on eliminating and preventing school problems;
- parents performed essential functions both in the school activities and at the decision-making level;
- the needs of students were dealt with thoroughly, and avoiding any possible inequalities between them;
- the educational community as a whole was part of the solution, favoring a better climate, more positive attitudes, better results, and an improved coexistence.

The team headed by Comer put this program to the test again in the public primary education schools of Benton Harbor (Michigan) from 1982–1983 to 1985–1986 academic years. Once more, the results were highly satisfactory, with an improvement not only of academic results but also of the climate (absenteeism, punishments, etc.) existing in those school centers (Haynes, Comer, & Hamilton-Lee, 1988). Without being the panacea and without actually reaching the status of a learning community, as we know it at present, Comer's SDP turned out to be the seed from which this model was going to originate. He proved with these two experiences that a transformation or change toward an approach based on communities or organized systems, without inequalities and ensuring the cooperation of all the groups involved, guaranteed optimum results, both academically and in terms of coexistence. Nevertheless, and despite the excellent results obtained, it did not imply a reformulation in the educational world, where work mostly continued to be carried out under a model which saw the student as the instruction unit (Roth & Lee, 2006).

Two more projects that acted as precursors of learning communities stood out in the 1980s. On the one hand, from the Centre of Educational Research of Stanford (CERAS), directed by Henry Levin, the program Accelerated Schools (Levin, 1998) was implemented in 1986 with remarkably good results in two marginal San Francisco Bay schools. On the other hand, Roger Slavin developed one year later, in 1987, his program Success for All, where more prominence was given to academic and social resources (Slavin, 1996) within that transforming mission of education shared by the three aforementioned programs.

When this whole movement that served as a precursor to learning communities emerged in the United States, precedents of something similar started to appear in some parts of Europe, including Spain (Morlà, 2015). An outstanding example can be found in the educational experience of Barcelona's La Verneda quarter, where a public center dedicated to the education of adults was organized as a community to deal with social inequalities, shaping free, democratic, participatory, and solidarity-oriented individuals (Flecha & Puigvert, 2002).

10.4 Learning Community Models

Five basic learning community models exist, according to Kellog (1999) and Tinto (2003), who make a proposal of classification for the university context which can also be extrapolated to the other educational levels. Similarly, despite being presented separately, these models are not mutually exclusive; elements belonging to several of them may coincide instead, which will depend on the characteristics and the environment of each educational center, which must adapt to them designing its own learning community model. Thus, these are the five possible models: (a) linked courses: each student is associated with two common classes or courses. One course is more theoretical and the other is a more applied class; (b) learning groups: learners study in several courses connected by a common link;

(c) groups of interest for first-year students, similar to the previous model but aimed only at first-year students; (d) federated learning communities: learners study in three or four related courses and in another course or seminar taught by a "learning master" who orients students; and (e) coordinated studies: this is a single course in which we actively work on an interdisciplinary theme.

Other more generic classifications are guided by different criteria, such as the environment where learning communities operate (Coll, 2001). We would thus speak about: (a) classroom-based learning communities: their scope of action is limited to the classroom; (b) school-based learning communities: their scope of action comprises the educational center or institution (Hill, Pettit, & Dawson, 1995; Shapiro, & Levine, 1999); (c) community-based learning communities; community learning networks; learning cities, learning towns, and learning regions: they refer to a city, a region, or territory with a variable extension where a group of people reside (Yarnit, 2000); and (d) virtual learning communities: virtual communities based on digital communication networks.

10.5 Foundations for the Development of Learning Communities

10.5.1 Pedagogical Principles

Regardless of their typology, all learning communities have a series of premises or principles which identify them and which can be summarized here from a variety of contributions (Flecha & Puigvert, 2002; Coll, 2001; Bielaczyc & Collins, 1999; Wilson, Ludwig-Hardman, Thornam & Dunlap, 2004; Bonk, C. J., Wisher & Nigrelli, 2004):

- Building a collective type of knowledge which serves as a model and a source of support for individual learning processes.
- The "learning to learn" competence is set in motion to make learning more effective and long-lasting.
- There are no barriers to learning: formal, nonformal, and informal learning are included.
- Learning conceived as something which takes place throughout life.
- A bet is made on educational equality, breaking the barriers and gaps that may exist. Education develops between peers and leadership is shared.
- It is considered an alternative to the current educational model in which education

—understood in its broadest sense—is transformed. This means changes in the spaces, curricula, functions, and roles of all the educational agents involved who, in turn, must commit themselves to this redefinition.
- A feeling of community or group membership arises which favors not only a better coexistence but also higher levels or self-esteem and security.

10.5.2 Stages in the Implementation of a Learning Community

A number of premises should be considered when it comes to designing and implementing a learning community model in any of its formats. Firstly, we must know the current situation of education, of the classroom, of the center, the territory, or the reality about to be transformed. This sensitization stage helps to identify everything that is not working and that we consequently want to change. It also becomes necessary to suggest what we would like to achieve, where we want to go, or which goals and approaches we would like to reach and embrace. Both reflections need to be made: where we stand; and what course we want to take; since, although the first issue may prove discouraging, the second one must be exciting enough for us to be able to go ahead.

In a second stage, the agents involved, the teaching staff, the educational administration, etc.—with the support of the whole educational community—will have to approve and undertake to successfully implement the learning community model. A third stage would envisage a joint design by all the sectors concerned of the specific learning community that they want to put into practice, which will necessarily be dictated by the type of learning sought for students.

The last stage would consist in setting the project in motion, which includes carrying out assessments and possible corrections during its development. This ongoing analysis and monitoring of the learning community's operation will permit to introduce improvements, as well as to customize the process for the purpose of enhancing the quality of this transformation.

Teachers' specific training is of paramount importance throughout these stages. This training will have to be guided and performed within the center or scope of action of the community in question (Flecha & Puigvert, 2004) and must be oriented toward the community's main concept and goal, i.e., where we want to go; and what we would like to achieve. It additionally needs to be approached on the basis of the teachers' already existing knowledge, beliefs, and attitudes, which must be contemplated—and modified if necessary—; otherwise, no real change will be accomplished (Van Driel, Beijaard & Verloop, 2001).

10.5.3 Virtual Learning Communities

The term "virtual learning community" comprises other terms with which this type of community has come to be identified: e.g., online learning community, online community, Internet community, digital community, or cybercommunity (Cabero, 2006).

As already explained earlier, we speak about virtual learning communities when ICTs are used to shape communication and information-exchange networks aimed at promoting learning. The main objective sought by these communities is a collective construction of knowledge by means of shared learning thanks to computer networks, which use synchronous or asynchronous means through the exchange of text, image, and sound.

Lots of technology and tools are available to these networks, and the range of resources never stops growing, thus offering new possibilities: synchronous resources (instant messaging, apps for mobiles, chats, etc.); asynchronous resources (forums, message boards, mailing lists, etc.), blogs, collaboration (wikis, shared documents, etc.), as well as social networks and social learning, amongst other things.

The same as in the case of learning communities, a variety of classifications can be found when it comes to virtual learning communities. By way of example, a first classification draws a distinction between two types of virtual learning communities:

- e-learning communities: groups of people who interact and communicate with one another exclusively using technology;
- blended learning communities: where interaction and communication take place both online and face to face.

According to the goal that they pursue, virtual learning communities can be classified (Cabero, 2006) into:

- Learning communities strictly speaking: they are designed to achieve a collective learning of the group who shapes them.
- Communities of practice: communities of individuals with roles aimed at continuous action and interaction, always seeking group support and strengthening. Shared reflections are made on practical experiences.

- Research communities: when their main goal is to set in motion joint and cooperative research projects between community members.
- Innovation communities: in this case, the aim sought consists in generating cooperative innovation processes.

Another classification based on the purposes of such communities establishes three categories (Van Riel 2004), namely: (a) knowledge-based communities; (b) practice-based communities; and (c) task-based communities.

On the whole, these virtual communities bring benefits (Gannon-Leary & Fontainha, 2007), amongst which stand out the following: an exchange of knowledge and learning; the synergies created; a feeling of connection, an improved learning environment; continuous and unlimited interactions; a mutual information exchange; and a fluid knowledge development. Barriers or difficulties similarly exist for their development: resistance to— and fear of—change; nonactive or nonactively-involved participants; participants with hidden identities; the use of an excessively technical language; and message misinterpretations due to the absence of body language; to which must be added that quite a high utilization of ICTs and a familiarization with them is required.

10.5.4 Professional Learning Communities

When a group of professionals get together to form a community with the aim of learning, we speak about professional learning communities. This approach is opposed to the one followed in many centers, where the teacher adopts isolated ways of working, without communicating with other teachers of the same center—or of others—as far as the educational practices developed by them are concerned. As seen *supra*, projects focused on transforming school have arisen which served to verify that, when teachers work as a team in a coordinated manner, inside open classrooms and interacting, their morale and motivation improves, as does the way in which they perform their function, and ultimately, their students' learning (DuFour, 2004).

In general, these appear as the main characteristics of professional learning communities (Hord, 2008):

- Shared vision and common goals
- Shared, solidarity-based leadership. Active involvement in all the proposed activities, in decision-making, and in everything that concerns the community.

- Structural as well as relational support. Working collaboratively allows for the establishment of trust relationships and encourages positive attitudes of respect and value between community members. This also benefits the times, spaces and resources to meet, to study, to work, and to learn.
- Collective learning and its application. This way of learning by the community facilitates improvements in their own praxis and, with it, the possibility to identify which actions are being successful and which are not, for the purpose of doing something about it or to introduce improvements.
- Shared personal practice. These communities achieve maximum effectiveness levels when their members are able to embrace teamwork, sharing groups or classes, and where teachers can observe their colleagues while teaching, take notes, and share remarks.

The establishment of professional learning communities in an educational context results in improved student learning (Siguroardóttir, 2010). Nevertheless, the same as every community, it requires a permanent monitoring, assessment, and analysis through which improvements can be made and its proper operation ensured for the sake of learning, both of its members and of students.

10.6 Conclusions

As seen earlier, learning communities are a complex world, with a wide variety of possibilities as well as peculiarities when it comes to their implementation. What began in the United States as a movement which revolved around projects aimed at a pedagogical transformation of the educational world in the 1980s, has evolved until the present day to become a transgressing model which is revolutionizing the educational world.

Now then, despite the verified benefits and successes brought by the establishment of learning communities as an educational transformational model—to which numerous publications have referred (Cross, 1998; Stassen, 2003; Graham, 2007; Weiss, Visher, Weissman & Wathington, 2015)—we must admit that it neither constitutes a panacea nor is exempt from criticism. Evidence has already been provided about the fact that the creation of a learning community in an educational center with an already existing teaching staff requires training and several implementation stages which need to be strictly followed. Hence, why overcoming teachers' fears and resistance to change represents a real challenge. Likewise, difficulties increase if we speak about other groups of agents involved, such as students, families, or even teachers of centers with studies previous or subsequent to those of the center in question, since not only higher levels of prejudice must be overcome but also more degrees of availability and willingness have to be faced, amongst other problems.

Despite these possible difficulties, we have also checked their high effectiveness together with their great success wherever this learning community models have been applied. They consequently become a guaranteed recipe for success in the transformation of the educational system, providing an improvement, not only in students' learning—and a better coexistence—but also progress is made in the learning of teachers, families, and the other members of the educational community.

This model can be implemented in any educational stage or level, and it has been enriched and improved thanks to ICT utilization. The so-called virtual learning communities offer endless advantages which generate new ways of building knowledge and of learning. Face- to-face teaching is no longer necessary; neither do we have to establish fixed spaces or times, needing to fix the attention in the right and specific moment. Instead, we can learn from anywhere and at any time, repeating contents if that is deemed appropriate, etc. Through this, the collaborative learning generated in virtual learning communities can achieve a higher motivation and response of those experiencing it.

Professional learning communities likewise permit to break the barriers of teachers isolated in their own classroom. Teamwork becomes the rule; teachers from various levels, stages, centers, regions, and countries are coordinated. And all of this allows for a personal and professional enrichment that goes beyond the individual, as it positively and directly affects their teaching function and, accordingly, their students' learning.

Acknowledgments

The present paper is framed within the Research Group "EDUTIC-ADEI" (Ref.: Vigrob-039) and ICE's Program of Research in University Teaching (REDES-I3CE-2017-4131), both of them from the University of Alicante; within the project Higher Cooperative Research Institute [IVITRA] (Ref.: ISIC/2012/022; http://www.ivitra.ua.es) and in the project "Second Round Art and Fight at Secondary School in Cinema and Audiovisuals" (UV-SFPIE_GER18-848892) from the University of Valencia.

11

Lifelong Education

María R. Belando-Montoro[1] and María Naranjo Crespo[2]

[1]Complutense University of Madrid, Spain
[2]University of Extremadura, Spain
E-mail: mbelando@edu.ucm.es; mafernandezs@unex.es

11.1 Introduction: On the Institutional Recognition of Lifelong Learning

Lifelong learning continues today, after several decades of international recognition and institutional support, a subject that is currently topical at all educational levels and areas. In fact, it is considered one of the main objectives of the supranational agencies and of the various national educational administrations. At the World Forum on Education organized by the UNESCO, it was proposed as an objective for the year 2030. And one year earlier, UNESCO (2014) in its Strategy 2014–2021, and in the framework of its general mission to contribute to peace and sustainable development, presents three strategic objectives. The first one is "to help develop and strengthen educational systems so that they provide learning opportunities throughout life."

After more than four decades since the publication of Report Faure UNESCO on learning throughout life, and more than two decades of transcendent and still indispensable reference, the Delors Report, this objective is still very much present, and still remains the central axis of world education policies.

Also at the European level lifelong learning has achieved a remarkable role. In October 1995, the European Parliament and the Council declared 1996 the "European Year of Permanent Education and Training." A few years later, at the European Council in Barcelona, it was asserted that lifelong learning has become indispensable to the knowledge society

(Consejo Europeo, 2002). These are just two examples of the intentions that some years ago to the EU were made explicit, underscoring the importance of lifelong learning for all European citizens.

This importance has also been recognized in Spain through educational legislation. Thus, article 5 of the Organic Law 2/2006, of May 3, of Education is entirely devoted to lifelong learning, maintaining the validity in the Organic Law 8/2013, of December 9, for the Improvement of Educational Quality.

After this recognition of the scope of lifelong learning, an approach to the description of this expression, and its characteristics, is necessary to help dispel any doubt about its delimitation and detail its content.

11.2 Concept and Components of Learning Throughout Life

From the organisms of the EU has prioritized the learning throughout life, especially in the decade of the 1990s of the 20th century, and in the first years of the 21st century. One of the fundamental documents of this period is the communication "Making a European area of lifelong learning a reality" (Consejo Europeo, 2002), in which the issue of employment was prioritized.

However, lifelong learning should not be inculcated only in the workplace. Consistent with this idea, the concept has also been extended to other areas, including four objectives: *personal fulfillment, active citizenship, social integration, and employability and adaptability.*

The evolution experienced by the concept of lifelong learning, in addition to the specification of specific objectives, principles, and strategies (Figure 11.1), responds to the impulse that from the Feira European Council (Consejo Europeo, 2000), and even earlier (October 23, 1995 the European Parliament and the Council declared 1996 as the "European Year of Permanent Education and Training"), has been provided for lifelong learning: motivated impulses, among others, for the consequences of globalization, demographic change, digital technology, and the deterioration of the environment.

According to this context, the Council Conclusions of 12 May 2009 on a strategic framework for European cooperation in the field of education and training expressed the need for this cooperation "to be placed in the context of a strategic framework that covers education and training systems as a whole, within a perspective of lifelong learning" (Consejo de la Unión Europea, 2009). And even more, it is argued that lifelong learning must

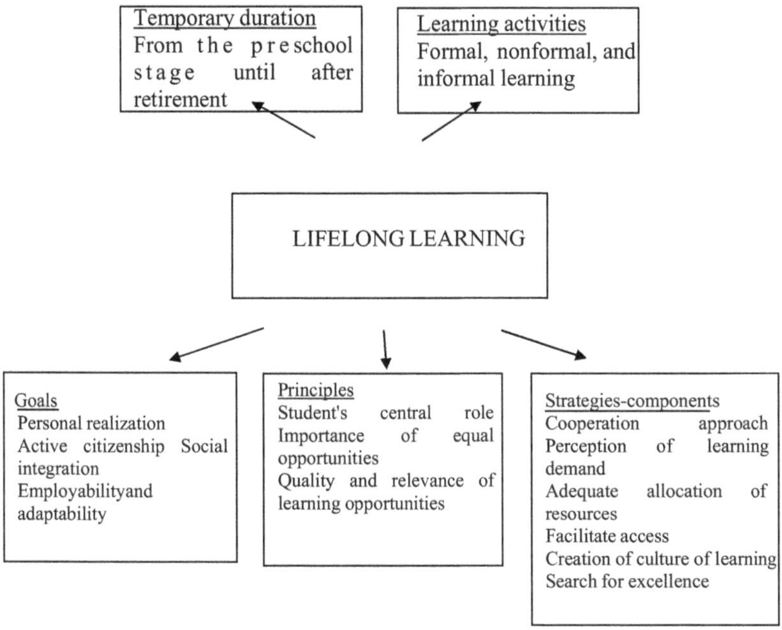

Figure 11.1 Elements that characterize lifelong learning (Comisión de las Comunidades Europeas, 2001).

be considered a fundamental principle that underpins the entire framework, which will encompass learning in all contexts and at all levels. Also, the latest evidence of prioritization of permanent learning throughout life on the European agenda can be found in the communication entitled "Building a stronger Europe: the role of youth policies, education and culture" November 19, 2018, which states that one of the priorities of the European agenda for the year 2019 will be "to work toward inclusive approaches based on lifelong learning and driven for innovation for education and training" (Comisión de las Comunidades Europeas, 2018).

Returning to the theme of objectives, if those who define themselves as involved in the concept of permanent learning extend the previous conception of it, alluding to personal, civic, and social dimensions, elsewhere in the Communication, reaffirms the support for this new definition to specify the two objectives of the European area of lifelong learning:

First objective: To train citizens so that they can face the challenges of the knowledge society, moving freely between learning environments, jobs, regions, and different countries in order to learn.

Table 11.1 Linking the priorities of action to the components of the strategies (Comisión de las Comunidades Europeas, 2018).

1. Valorization of learning Information	• Creation of a learning culture • Facilitate access to learning opportunities • Search for excellence
2. Information, orientation, and Advice	• Facilitate access to learning opportunities • Creation of a culture of learning • Work in cooperation
3. Invest time and money in learning	• Adequate allocation of resources • Facilitate access to learning opportunities • Search for excellence
4. Bring learning opportunities closer	• Creation of a learning culture • Work in cooperation • Perception of learning demand
5. Combining basic skills	• Perception of learning demand • Creation of a learning culture • Facilitate access to learning opportunities
6. Innovative pedagogy	• Perception of learning demand • Work in cooperation • Search for excellence

Second objective: To fulfill the goals and ambitions of prosperity, integration, tolerance, and democratization that the EU and the candidate countries possess.

The achievement of these intentions requires concerted action in accordance with mutually agreed priorities associated with the six key messages that formed the core of the Memorandum on Lifelong Learning (Comisión de las Comunidades Europeas, 2000), and on which a Europe-wide consultation was based (Table 1).

In response to the recommendations issued by the EU agencies, as well as other supranational organizations, Spanish educational legislation has been incorporating, recognizing, and developing the importance of lifelong learning in recent years.

Of special significance is that article 5 of the Organic Law 2/2006, of May 3, of Education, is entirely dedicated to lifelong learning, maintaining the validity in the Organic Law 8/2013, of December 9, for the Improvement of Educational Quality.

Two years after the promulgation of the LOMCE, the Ministry of Education, Culture and Sports (Ministerio de Educación, Cultura y Deporte, 2015) published the "Strategic Lifelong Learning Plan" with the aim of promoting education and training throughout the life of the Spanish citizens.

However, despite the recognition of this right, and the support of educational administrations, the data reveal that participation in training activities after 25 years does not even reach 11%. If in the European context the data show that in 2017 only 10, 9% of the population between 25 and 64 years old had participated in some training course in the last four weeks (EUROSTAT, 2019), the Spanish figures do not yield positive results, nor are they very encouraging (from 11, 2% in the year 2012 to 9.9% in 2017) (EUROSTAT, 2019). In the case of the adult population, this low participation rate is largely due to lack of time, since between work and family obligations one hardly has time to participate in formative processes.

11.3 Lifelong Learning Methodology: A Proposal for the University Classroom

According to Puig & Palos (Puig & Palos, 2006), service learning is defined as "an educational proposal that combines learning processes and community service in a single well-articulated project, in which participants are trained to work on real needs of the environment with the objective of improving it." To deepen this definition, the essential characteristics of the educational proposals designed and developed based on the learning-service methodology are presented hereunder (Batller, 2011; Moliner, 2010; Mayor, 2017; Mendía, 2012; Puig & Palos, 2016, Pérez & Ochoa, 2017):

- It is a profound approach to *learning*, in that it develops teaching-learning processes oriented toward the integral formation of the individual.
- It is *service*, since it aspires to identify the real needs of society with the aim of improving them.
- It is applicable to any educational context, both in the field of formal and nonformal education, and at any age and educational level.
- It is based on the principles of experiential and reflexive pedagogy.
- The entire educational community is involved in the teaching-learning processes.

– Processes of individual transformation are developed that in turn provoke processes of social and community transformation.

Although the first experiences of learning service arose in the decade of the 1920s of the last century—based on the pedagogical approaches of John Dewey, William James, as well as the educational plans of the pedagogues of the New School movement, Paulo Freire or Célestin Freinet (9, 11, 15)—currently is still considered as an innovative educational proposal. In this sense, then, the basis for the full expression of the potential of this methodology are described to address the main objectives of lifelong learning discussed earlier: personal fulfillment, active citizenship, social integration, and employability and adaptability (3).

Educational projects designed and developed under the paradigms of this methodology, start from the identification of the needs of the environment with the aim of developing actions to improve them, thus promoting active participation in social, political, and community life of the students, and the rest of the educational community; as well as addressing the improvement of social, civic, and political competence. In this sense, these projects favor the development of the following aspects in the student body (Batller, 2011; Moliner, 2010; Mayor, 2017; Mendía, 2012; Pérez & Ochoa, 2017; Puig & Palos, 2016; Vila, Castro, Barreiro & Losada, 2016):

– Responsibility;
– Social and political commitment;
– Search for the common good;
– Feeling of belonging to the community;
– Empowerment;
– Analytical, associative, and reflective capacity;
– Critical thinking;
– Ethical, social awareness, and democratic values;
– Moral sensitivity;
– Interest in social issues;
– Understanding the sociocultural and economic context;
– Solidarity;
– Social entrepreneurship;
– Respect for human rights;
– Empathy;
– Respect for cultural differences;
– Moral integrity.

Landing in the field of higher education and in line with the manifest need to move toward an effective model of university social responsibility, the development of educational proposals located within the paradigms of the methodology of learning service can contribute–moving along the lines of the development of individual aspects mentioned earlier—to build one, active, caring, responsible, critical layer of social awareness, and meet new needs and challenges of the 21st-century societies toward inclusive citizenship; and actively involved in the construction of a real welfare society, which highlights aspects such as collaborative relations, social cohesion, the continuous improvement of the community and social, personal, educational, and labor inclusion.

Likewise, although this methodology does not necessarily require the inclusion of ICTs in the development of their educational processes, these will undoubtedly play an important role as facilitators of learning. In this regard, the LOMCE states that "Information and Communication Technologies will be a key tool in teacher training and in the learning process of citizens throughout their lives, by enabling them to make training compatible with personal obligations, and/or, in the same way, they will be in the management of processes" (Ley Orgánica, 2013).

In this sense, and within the framework of the educational processes developed under the learning-service methodology, a possible proposal to develop in the university classroom would be the construction of ICT resources to meet the socioeducational needs identified in the environment; from one's own faculty to colleges, institutes, day centers, etc.

For this, in the first place, the group class will be divided into small work groups (no more than six students per group), each of which will choose a space of educational and/or social character from the environment close to the faculty to develop their proposal.

Secondly, after selecting the space, one should start the process of identifying the socioeducational needs of the environment. In this sense, and with the aim of providing the process with a descriptive and interpretative approach oriented to the construction of meanings, it is advisable to adopt a qualitative ethnographic research mode, using instruments such as the ethnographic interview, or the research group discussions.

In the third place, a search and analysis of websites of Spanish and international institutions on lifelong learning will be carried out in relation to each of the objectives of lifelong learning: personal fulfillment, active citizenship, social integration, and employability and adaptability

(Comisión de las Comunidades Europeas, 2001). The analysis of the webs will attend the objectives, contents, activities, scope, and evaluation in which the results of the plans, programs, projects, or activities carried out in the last five years by each of the institutions are detailed. Likewise, in this phase we will continue with a search of ICT resources to help achieve the four objectives of lifelong learning, as well as to improve the needs identified in the environment. For the selection of these ICT resources, pedagogical criteria will be taken into account. In addition, they must be free resources, easy to use, and must have an app for mobile devices.

Finally, the process will culminate with the creation of a blog by the students, which will reflect the experiences developed by each of the working groups, as well as the development of a MOOC by the teachers on lifelong learning, ICTs, and inclusion.

The evaluation of this experience will be formative (participatory, consensual, and focused on learning) and will use a reflective diary, observation sheets, and a questionnaire. These instruments will allow the assessment of both issues related to learning processes, as well as the content and skills acquired, and the achievement of the objectives set.

In this way, through the development of this type of proposal, a view will be adopted that places ICTs as facilitating tools for the development of inclusive educational processes, and aimed at achieving the principle of education for all.

11.4 Conclusion

The exposed data lead us to a concern about the low participation in continuous training activities. A circumstance that, as has been mentioned in previous sections, is happening in all the member states of the EU, except for some exceptions, such as Denmark, Sweden, and Finland (EUROSTAT, 2019). At the heart of this problem are probably the consideration that learning is not attractive, and additionally that there are other activities that arouse more interest. Hence, one of the questions on which one might reflect in this regard is on what measures could be taken to motivate learning, and also on informing and facilitating access to learning. In any case, the fact that the development of the free online courses that are being offered on different platforms for some years now, and among whose advantages highlights flexibility, cost savings, and autonomy, has not been assumed in this matter an increase in interest and in the formation of the adult population.

Although, this is only one of the problems that are preventing further development of learning throughout life, as, as we have seen, lifelong learning is not only what the adult population performs.

What is learning throughout life? The breadth of the concept would require more space than these pages offer, but a definition that synthesizes the essentials of this paradigm may be a good starting point: lifelong learning is a global way of understanding learning, a principle on which the organization of a structure and learning contents is based, and supposes a project that includes all the possibilities of training in any field of knowledge, and at any moment of a person's life.

However, it would be necessary to clarify certain essential elements, certain characteristics that make learning throughout life a priority objective for human development. First, the active role of the person who learns through processes of reflection and assimilation, in addition, the person must be able to learn on their own. Secondly, the integration of formal, nonformal, and informal learning; and, thirdly, objectives aimed at personal fulfillment and, at the same time, at their civic, social, cultural, and labor development.

In the context of this concept, a line of doubted interest for reflection and research on the development of lifelong learning is related to the motivation toward learning, and in a sense of living; wanting a life of quality that expresses a deep satisfaction of the mere fact of life itself.

12

ICT Teacher Training and Disability

José María Fernández Batanero and José Fernández Cerero

University of Seville, Spain
E-mail: batanero@us.es; jfercerero@gmail.com

12.1 Introduction

In the last 30 years there has been a profound change worldwide, particularly in terms of scientific developments and social changes. The impact on society in general and on education in particular, is increasingly visible, being in a permanent state of transformation and improvement due to the dizzying proliferation of ICT. The educational environments can be opened to the world thanks to these new resources, and can interact with other realities, thus opening up a new space of access and production of learning, contributing at the same time to the elimination of the barriers that prevent the approach of all the people to the educational fact. This interest in equality and equity is evident in all international initiatives in the last decade, as an example, one can point to the project "Taking advantage of ICT to achieve the 2030

Education goals" of the UNESCO-Weidong Group Fund, that will continue to help to put into practice this educational idea for the next four years allowing participating member states to take advantage of the potential of ICTs to achieve SDG 4 by 2030. Also at its General Assembly on December 13, 2006, the United Nations adopted the resolution drafted by the International Convention on the Rights of Persons with Disabilities, which asserts and states that to ratify, it must undertake or promote research and development of ICT accessible to people with disabilities, as well as their availability and use, including specific technical devices created to improve the daily life of this collective.

12.2 Conceptualizations: ICT and Disability

The concept of technology has been referred to by specialized bibliography from different perspectives. Thus, from a technical point of view, authors such as Haag, Cummings & McCubbrey (2004), considered that information technologies were composed of "any tool based on computers and that people who are used to work with information, support the information and process information needs." From an institutional perspective, the OECD (2002) defined ICT as "those devices that capture, transmit and deploy electronic data and information and that support the growth and economic development of the manufacturing industry and services." From an educational perspective, authors such as Luque Parra & Rodríguez Infante (2009) understand ICT applied to education as any means, resource, tool, technique, or device that favors and develops information, communication, and knowledge, a definition that entails a marked practical and applied character, within the field and educational system, so it should be considered also, as a didactic support for learning, an element for cooperative work, and also as an element of management and administration. On the other hand, Hegarty, Bostock & Collins (2000) presented a definition of ICT intervention for special needs from a critical and a technical perspective, i.e., critical aspects (i.e., attitudes and self-confidence) and technical aspects. We carried out a classification of four categories: cognitive difficulties, emotional difficulties, physical disability, and sensory disability. The definition highlights the dimensions of activity and participation. In this context, ICT support provides opportunities to increase the performance of activities and/or opportunities to increase participation.

For Toledo, Sánchez & Gutiérrez (2013) ICT facilitates access to all types of information more easily and comfortably; they favor the autonomy of the students, adapting to their needs and demands; they contribute to the synchronous and asynchronous communication of these students with their classmates and faculty; they help to adapt to the environment; improve cognitive development thanks to the activities that can be done; save time for the acquisition of skills and abilities; favor the diagnosis of the student; support a multisensory communication and training model; to promote an individualized formation favoring the advance of the student at his/her own pace, which is of extreme importance for these subjects; favor the development of autonomy and independence; avoid marginalization, the digital divide, which introduce possibilities to people who are unable to use the tools of development of the knowledge society; facilitate the sociolabor insertion of the student with specific difficulties; provide leisure time; save time for acquiring skills and abilities; the exercises that the students must carry out can be executed and repeated with minimum efforts so that the students acquire the competences, attitudes and capacities; encourage the approach of these people to the scientific and cultural world; or that they can be excellent simulators.

Regarding the term disability [1] we must say that it varies according to the researchers and the context. From the perspective of the International Classification of Functioning, Disability and Health (CIFDS, WHO, 2001), disability is understood as the circumstance of negative aspects of the individual's interaction and contextual factors, limitations of activity, and restrictions in general participation. In the traditional medical model, a "disability" is defined as any form of impairment or limitation in the normal functioning of an individual, so that "impairment" implies a reduction or weakening of normal functioning, and "limitation" implies seriously reducing daily activity.

12.3 Research in ICT and Disability

In the pedagogical field, learning with ICT to support people with disabilities has been the subject of research for several decades, but it has been in the last 10 years that it has become an important part to support the learning of students with disabilities. A quick search with the terms ICT and disability in Google Scholar revealed 139 results between 1988 and 1998, 3150 from 1998 to 2008, and a staggering 15900 from 2008 to 2018. It is likely that not all of these documents are directly related to the use of ICT. To support the learning of people with disabilities, however, the large number of contributions identified suggests that there is a growing literature that arouses interest in the subject. In this regard, in the case of Europe, in 2010 the initial results of the project "European Research Agenda for Equality of Disability" were published, which encourages the participation of civil society organizations to participate in research with academic institutions, where technology plays a predominant role (Priestley, Waddington & Bessozi, 2010). In this paper, the authors focused on ICTs to support learning in different areas (access to ICT, teaching methods and learning, development and testing ICT solutions, reviews, evaluative articles on inclusion, social development, and behavioral documents, use of ICT as mediators to interact, digital games, etc.).

In the international arena, scientific studies have been carried out that have already been in charge of bibliographically reviewing the support of ICT for people with special educational needs (SEN). Fitzgerald, Koury & Mitchem (2008), conducted a review of the literature, from 1987 to 1996, on studies related to computer-mediated instruction in the learning of students with mild and moderate disabilities. On the other hand, Pennington (2010) reviewed the research carried out between 1997 and 2008 using computer-assisted instruction to teach academic skills to students with autism. While Fitzgerald, Koury & Mitchem (2008) analyzed their findings in terms of teaching methodologies, changes, reforms, and inclusion, Pennington's review (2010) focused on the design and effectiveness of computer-assisted instruction, as well as the methodology used in the studies, in addition to identifying various types of learning.

In a more recent article by Liu, Wu & Chen (2013), ICT trends were revised to support students with SEN, where they provided an exhaustive analysis of 26 studies published in journals indexed in the last five years (2008–2012), focusing on the research objectives, the methodology used, and the results.

Istenic & Bagon (2014) presented the results of a content analysis of all articles published from 1970 to 2011 in seven educational technology journals indexed in Web of Science (WOS). This study highlighted the scarcity of jobs related to ICT and disability, and that these are published especially in educational technology journals. At the same time, it is highlighted that the question of the potential of learning supported by ICT for the process of inclusion of people with SEN has not been sufficiently explored. The study ended by stating that documents on this topic began to appear in 2001, and during the period 1970–2011 only 17 articles were published.

On the other hand, Hersh (2017) produced a classification framework for learning technologies based on inclusive ICT and ICT-based learning technologies for people with disabilities, covering general learning, and assistance technologies. The classification is important because it helps to structure and understand the field, determine good practices, and facilitate the combination of technologies with students.

In short, we can see how the studies on ICT and disability are being linked thematically with the importance of ICT with students with disabilities, the environment, and their use. Hence, we emphasize that there is a line of research that is not very developed, such as the professional development of teachers, to prepare them for the use of ICT and educational inclusion, i.e., to provide them with skills and competences for their own professional learning and for teaching. We must not forget that if we intend to obtain good practices with ICT, teacher training plays a key role, since innovation and learning depends on it.

12.4 Teacher Training in ICT and Disability

The educational policies of different countries are rushing to incorporate the development of "digital competence" as one of their basic objectives, and the laws of education include in their preambles references to the importance of information and communication technologies. In this sense, teacher training has been affected by the urgency of technically training practicing professionals. When a first level of learning of the management of devices and programs has been overcome, it has been to focus on the didactic dimension, on analyzing the potential of new media to favor the teaching-learning processes, but on very few occasions has it been possible to underline teacher training for media education.

There are many studies that have been carried out regarding digital teaching competences in general, but very limited when those digital teaching competences imply support for the learning of students with disabilities or functional diversity. Along these lines, studies have been carried out that inform both the development of competences in the initial training (Istenic, 2010), and permanent (Fernández Batanero, Cabero & López Meneses, 2018) to design learning environments that respond to individual needs.

At a general level, in relation to digital competences, the studies carried out with respect to their training for ICT management suggest that teachers have positive attitudes toward them, but feel insecure for their incorporation into the teaching-learning process, and not so much from a technological point of view, but rather from a didactic and methodological perspective (Prendes & Gutiérrez, 2013; Suárez et al., 2013). This explains the low variability of technological materials that teachers use with students in their professional activity (Ferrandis, Grau & Fortes, 2010).

In the case of their training for the use of ICT applied to subjects with functional diversity due to disability, the first thing to point out is the strong marginality of these works in the European and Latin American context, as observed in those that address the problem of ICT training for teachers, and the skills they need to use ICT with people with disabilities (Molina et al., 2012, Rosario & Vázquez, 2012, Terigi, 2013, Rangel & Peñalosa, 2013, Ortiz et al., 2014). On the other hand, the few studies conducted point to the lack of training and knowledge that teachers have regarding the different types of technologies that can be used, the possibilities they offer, and the various functions for which they can be used (Fernández Batanero, Román & El Homrrani, 2017) .

In a recent study (Fernández Batanero, Cabero & López Meneses, 2018) that sought to know the degree of knowledge and training of Spanish primary education teachers regarding the application of ICT to support the learning of students with functional diversity, the lack was highlighted of training and qualification, for the use and application of ICT in students with disabilities. Teachers show even at advanced levels that ICT is a useful resource for the education of people with disabilities. Now technical management and the educational outlook in general, use computers and the Internet, reverberated in the knowledge that they possess, testify to the application of these technologies for people with disabilities.

The inappropriate use that teachers make of ICT in the classroom increases if, in addition, we take into account the barriers that hinder teacher training. Diverse are the investigations that show the factors that prevent the implantation of the ICT in the scholastic centers. Among them, we highlight those that refer to extrinsic barriers (lack of time, resources, and funding to carry out the training) and intrinsic barriers (attributed to the attitude of the teaching staff). A study in the Spanish context revealed that the main barriers that hinder the implementation of ICT training activities and disability in most of the autonomous communities are determined in the first place by intrinsic factors such as economic, time availability, and attitude of the teaching staff. It also shows that the lack of training and confidence of teachers in ICT is crucial for their commitment to them, which implies a direct relationship with quality and quantity (Fernández Batanero, Reyes & El Homran, 2018).

12.5 Conclusions

The lack of training clearly denotes the need to promote educational policies on initial and ongoing teacher training in the application of ICT for people with disabilities. In this sense, we consider the need to adopt urgent measures so that teachers are trained and incorporate the technologies in their daily practice with people with different types of disabilities. Therefore, an initial training is required at universities and higher education centers that includes knowledge of materials, software, websites etc., that favors the application of strategies and adaptations, that take into account accessibility criteria to develop educational materials, and that facilitate the creation of learning environments according to the reality of the centers, the demands of the teaching staff and, mainly, the characteristics of the students. Teachers must receive conceptual training in relation to ICT within the educational context, how they transform and support the learning environment and social equality; all of this can help to change beliefs about ICTs and help the teacher to update and renew approaches in their teaching work, as well as to create their own content and educational resources, and changing their role as a repeater of the experiences and practices of others. The reality that we face in quotidian experience has more innovative and enriching technological tools that offer us a wide range of applications and adequate tools for this purpose.

In relation to lifelong learning, low training in digital teaching skills can make it difficult to fully include students with different types of disabilities, especially if it is the school's responsibility to create an analogue and digital scenography that facilitates the inclusion of the person. Consequently, educational centers and teacher training centers should promote this training, where we can highlight three ways: (a) the holding of courses, seminars, etc., (b) the putting into action of meetings in which they report to the teaching staff of all the activities that are published, where information is disseminated, knowledge related to the subject is shared, or "new practices" of incorporating ICT for students with diversity are made known; and (c) the incorporation of ICT for teacher training with training actions through the Internet, either in the e-learning or MOOC mode.

Finally, to say that the digital training of teachers should contribute to the integration of ICT in the classroom, not as an incorporation of new teaching resources (reinforcing the traditional models, unidirectional teaching), but an incorporation that brings educational innovation. That is, making "use of ICT" for pedagogical purposes, overcoming its use as a mere resource in the classroom. ICTs enable teachers to realize concretely that the activity of the student is the basis in the construction of knowledge.

12.6 Acknowledgments

This chapter is part of a larger research project funded within the framework of the State Plan for the Promotion of Scientific and Technical Research of Excellence 2013-2016 (DIFOTICYD EDU2016 75232-P).

13

The Use of Technologies for Supporting Diversity

Carmen María Hernández Garre,[1] María del Mar Fernández Martínez,[2] José Juan Carrión Martínez,[1] and Antonio Luque de la Rosa[1]

[1]University of Almería, Spain
[2]University of Huelva, Spain
E-mail: cmhgarre@ual.es; mar.fernandez@dstso.uhu.es; jcarrion@ual.es; aluque@ual.es

13.1 Introduction

The establishment of ICTs in our society has generated a series of transformations in the processes of teaching and learning, giving rise to an unprecedented revolution in the way in which knowledge is accessed (Cebrián, 2011). It is therefore necessary to use innovative resources that contribute toward the student becoming an active learner, which enhances his/her learning (Amar, 2013; Rivas & Rodríguez, 2015). In this regard, ICTs have become established as effective resources within the educational field.

Further, the use of these technologies allows for adaptation to the different learning patterns of students, taking into account the varied interests or needs that they present, thereby favoring the provision of support for this diversity by responding to the current educational reality of the classroom. Thus, educational inclusion represents an element that is key in the active participation of all students (Azorín & Arnaiz, 2013).

In this regard, ICTs are a key factor in the normalization of the socioeducational conditions experienced by students with special needs (Soto, 2007). Thus, adequate training of education professionals and the design of more accessible materials will be essential if these changes are to produce the maximum possible development of the potential of each of the students that make up the diversity of the classroom (Alba, 2012).

However, as noted by Cabero (1999), ICTs alone do not bring about changes even though their innovative characteristics can offer an effective and quality response to the diverse needs of students based on the practice of teaching for educational improvement, with commitment to inclusion being an aspect that affects all education professionals (Luque, 2012). Inclusive education requires teachers to assume responsibility for the transformation of organizational, curricular, and personal elements so that all students feel valued. This transformation also implies the involvement of the entire educational community to ensure that the principles of equality, equity, and social justice are upheld for all students (Arnaiz, 2012).

In contrast, the implementation of information and communication technologies has also generated new forms of social exclusion, which, according to Soto (2007), include the following:

- Absence of specific policies regarding digital inclusion.
- Difficulties in accessing technological infrastructures.
- Insufficient ICT training.

Based on these considerations, it is possible to question the role of technology as an instrument that promotes equal opportunities, given the potential difficulties faced by people with disabilities when accessing increasingly digitized knowledge. This could lead to technology producing a digital divide as opposed to being regarded as a facilitator of such knowledge (Abascal et al., 2015; Macdonald & Clayton, 2013). In fact, the professionals responsible for the design of these instruments should ensure access for all people regardless of their cognitive and physical abilities. In the case of children with disabilities, procuring adequate levels of accessibility and usability implies the need to consider their cognitive development, ability to verbalize, levels of abstract thinking, extroversion, ability to concentrate on an activity, and their previous experiences with the use of technology (Markopoulos & Bekker, 2002).

The review carried out by Escobar et al. (2016) highlights the various paradigms most representative of the participation of children with disabilities in the development of technology for the design and development of technological applications. Thus, contextual design involves the design of technological products according to an in-depth understanding of how users work, along with their needs. Design for all aims to ensure that all products cover and can be used by as many people as possible, and thus there is no single design solution (Persson et al., 2015). Participatory design is a methodology that starts with a more efficient technological design that engages users to participate in a more social way. Finally, inclusive design is understood as the design of products or services that are accessible and usable by as many people as reasonably possible, in a variety of situations, and with the widest possible scope without making adaptations or specialized designs.

Considering the above, the participation of children with disabilities in the field of technological development poses an ambitious challenge that should not only involve their participation for testing or identifying problems at any stage of the construction of the technological tool, but should also focus on more active participation in its design (Escobar et al., 2016).

In this regard, authors such as De Salces et al. (2014) and Kärnä et al. (2010) have proposed methodologies that encourage the collaboration between children with disabilities and R&D groups.

13.2 Digital Technology and Support for Diversity

13.2.1 Methods

13.2.1.1 Databases

The search for information was conducted with the appropriate rigor, and thus certain inclusion and exclusion criteria were proposed for the selection of documents.

The inclusion criteria that the studies had to meet were: (1) That they were carried out within the period 2014–2018; (2) The studies used quasi-experimental research methodologies, case studies, and theoretical reviews; and (3) The language used was Spanish or English.

The exclusion criterion was: (1) Studies that do not indicate the type of application used.

The search procedure used the following various sources of information: Google Scholar, WOS, Scopus, and Eric. In addition, several combinations of keywords were used, including special needs, functional diversity, information and communication technologies, ICT, technologies, diversity, technology, and diversity.

As a search criterion, the combination of keywords in the title, summary, and keywords was established, and the search process was carried out during the month of December 2018.

13.2.1.2 Descriptors

For the search of the documents several combinations of key words were used such as special needs, functional diversity, information and communication

Table 13.1 Results of the search using the information channels.

Information channel	Search results	
Language	Spanish	English
Google Scholar	757	1960
WOS	2	77
ERIC	0	446
SCOPUS	0	11

technologies, ICT, technologies, diversity, technology and diversity (Table 13.1).

13.2.2 Results

The relationship between ICTs and functional diversity due to disability represents an essential line of research within the framework of support for diversity and educational inclusion.

ICTs offer a wide range of possibilities that are capable of overcoming the shortcomings of conventional systems in the teaching-learning process and enhancing equal opportunities whilst at the same time these technologies are also considered to be a fundamental element in the development and enhancement of support for diversity (Fernández-Batanero, 2018; Cabero, Fernández-Batanero & Córdoba, 2016).

The use of ICTs offers a variety of advantages in inclusive classrooms (Romero, González & Lozano, 2018), including:

- The provision of help in overcoming the limitations caused by various sensory, motor, or intellectual disabilities
- The facilitation of more personalized attention to the students
- A contribution toward improved communication
- The promotion of student autonomy
- Time-saving
- Access to a multitude of information resources, which could enhance both free time and learning.

Taken together, all of the above serves to highlight the potential of ICTs to improve the quality of the teaching delivered to students with functional diversity. The selected studies demonstrate the importance of the use of technologies as a strategy for supporting students with diverse needs. Thus, some studies are focused on the potential of ICT in the teaching and learning of students with autism spectrum disorders (ASD) (Lozano et al., 2013) whilst others have shown the effects produced by the use of educational software in students with SEN (Orozco, Tejedor & Calvo, 2017).

However, there are a number of studies that indicate the need for preliminary and ongoing teacher training in digital competences aimed at supporting diversity, as well as the importance of improving study plans (Fernández-Batanero & Rodríguez-Martín, 2017; González & Gutiérrez, 2017; López & Ortega, 2017).

Educational inclusion implies the right of all students to have access to the relevant aids according to their needs, without discrimination, following the principle of equal opportunities. The change toward alternative methodological models necessarily involves the action of teachers who must have the training necessary for both planning and implementing innovative work with students who have SEN.

In this regard, the International University of Valencia (s.f.), and following Romero, González & Lozano (2018), recommends the following for favoring educational inclusion:

- The use of varied teaching methods and personalization of learning experiences.
- The promotion of the participation of parents in their children's school activities.
- The establishment of an active dialogue between the members of the educational community.
- The scheduling of activities (by the center) that require collaborative work with the various institutions and associations in the environment.
- The development of ties of coexistence and tolerance of the educational community through respect for diversity.
- The prevention of the exclusion of students by valuing diversity and interculturality.
- The provision of the necessary resources to address the needs that students may present.

In short, these practical recommendations aim to achieve a quality of education that responds to the diversity that is currently presented by students.

13.3 Conclusions

The effective inclusion of students with SEN requires a quality response from the educational centers that ensures the development of the potential of the students regardless of their needs.

It has been demonstrated that the use of ICTs offers a series of benefits for people with disabilities by providing multisensory stimulation, favoring attention, and reinforcing the ability to motivate. Similarly, the use of ICTs makes it possible for students to work independently (insofar as this is possible) and encourages the development of the capacity for self-monitoring, whist also adapting to personal characteristics, favoring integration. ICTs therefore constitute an active, versatile, flexible, and adaptable element of the learning process.

However, teacher training in information and communication technologies is essential to ensure the effectiveness of the teaching-learning model. Thus, one of the determining factors in the use of ICTs in education is the initial and ongoing training received by teachers (Ramírez, Cañedo & Clemente, 2012).

Another aspect to take into account is control in the use of ICTs as a teaching aid, ensuring that it must never serve as a substitute for the teacher. In addition, it is necessary to be aware of the possibility that the student might become dependent on the technologies.

A key factor in the use of ICT tools and applications is related to the knowledge of the characteristics of the people that are going to work with this technology and the objectives set out, given the heterogeneity of students with SEN.

As a response to the diversity of people with disabilities, there is an increasing presence of software, applications, and computer material that is useful for the areas in which they present the greatest difficulties such as social skills, communication, imagination, and recognition of emotions (Romero, González & Lozano, 2018).

In spite of the importance of integrating technology to encourage the progress of all students in the international arena, there are still relatively few studies that give special consideration to students with SEN (Wallace & Georgina, 2014).

Finally, as pointed out by Fernández-Batanero (2018), ICTs can contribute toward achieving higher quality teaching and learning that promotes social justice, provided that teacher training is adequate for the correct use of ICTs, with specified plans of ongoing development according to the current circumstances faced by the students.

14

Multiple Intelligences and ICT

Ana Marta Santamaría Rastrilla[1]**, Eva Ordóñez Olmedo**[1]**, and José Gómez Galán**[2]

[1]Santa Teresa de Jesús Catholic University of Ávila, Spain
[2]University of Extremadura, Spain, and Ana G. Méndez University, Puerto Rico-USA
E-mail: ana_rastrilla@yahoo.es; eva.ordonez@ucavila.es; jgomez@unex.es; jogomez@uagm.edu

14.1 Introduction: Concept and Types of Multiple Intelligences

There is a great difference between the conception of intelligence that we have today and that which existed a few decades ago. Howard Gardner's theory of multiple intelligences has mainly contributed to this discrepancy. Prior to his theory, intelligence was understood as a unique capacity measurable through IQ. From the milestone of Gardner's theory, intelligence is understood as composed of several abilities, being these autonomous.

Gardner established in his theory eight different intelligences, each one of these intelligences is present in all people with a different degree of development. And none of them is more important than the others, although there is always a tendency to think that intelligences such as mathematical-logical or linguistic are more important (Gardner, 2005). The division stipulated by Gardner is the following:

Intrapersonal intelligence: Can be defined as the ability to know oneself. To be aware of one's own strengths, as well as one's own weaknesses. This deep knowledge can be used to define the personal goals of each individual in a realistic way. Examples of people who possess this more developed intelligence are those who are reflective, reasonable, and usually have a good point of view, being able to give good advice to their neighbors.

Interpersonal intelligence: This capacity allows us to know and interact with the people around us. It is also related to the capacity of leadership, to organize, communicate, and solve conflicts. Good negotiators, those who enjoy working in groups and also those who have a great empathy can serve as an example of a high degree of development of this intelligence. This is a quality highly valued by companies today.

Linguistic/Verbal intelligence: This intelligence is related to communication in all its aspects, to handling and structuring the meanings and functions of words and language. It is knowing how to use language both to express oneself and to understand meanings. The syntax and phonetics as well as the semantics of language play an important role in this ability. Good speakers and politicians, writers, or those who are good at learning languages have a great development of this intelligence.

Logical-Mathematical intelligence: The capacities that mark this logical-mathematical intelligence are the handling of numbers with fluency, but also the ability to structure the problems to make deductions and to be able to solve them. Those who possess this intelligence stand out for their ability to analyze problems easily. People bestowed with this intelligence are those who also like numerical calculations and statistics. This is the case of mathematicians, scientists, economists, and computer scientists.

Physical, Kinetic/Corporal intelligence: This is the intelligence related to the physical parts of our body, which allows us to dominate our own body to express ourselves, train and compete. This implies coordination, balance, dexterity, strength, flexibility, and speed. This ability includes being able to unite body and mind to make optimal use of physical abilities. Not only sportsmen and dancers are examples of a development of this intelligence, but also people who are skilled in moving, playing instruments, DIY, or construction tasks.

Visual-Spatial intelligence: This intelligence is mainly defined by the ability to spatially represent ideas. Also, possessing the ability to use symbolic systems and sensitivity to color, shape, and space. A high degree of this intelligence is possessed by those who can imagine and solve spatial problems with ease and learn through vision, as is the case with painters.

Musical intelligence: This type of intelligence enables us to appreciate and transform musical forms and express ourselves with them. Identifying rhythm, tone, and timbre. This intelligence is reflected in musicians mainly, but also in people who are attracted to music.

Naturalistic intelligence: Nature, fauna, and flora are encompassed in this intelligence to the extent that people can understand them. It includes the urban and rural environment because it encompasses the skills of observation and hypothesis posing/test. Individuals who possess this intelligence stand out for their keen curiosity about natural phenomena and their love for animals. These include gardeners, ecologists, archaeologists, physicists, and chemists.

14.2 Teaching Models Based on Multiple Intelligences

The theory of multiple intelligences was initially put into practice in schools in the United States, through Gardner and his *Project Zero* collaborators. This is a research group from the Harvard School of Education, founded in 1967, whose aim is to apply multiple intelligences in the different educational stages. The following projects, taken from the literature presented by Muñz and Ayuso (2014), are noteworthy:

The Spectrum Project: Aimed at students in infant and primary education who are committed to individualized education. It was started in the 1980s to counteract educational practices with a rigid curriculum and evaluations that were essentially based on intelligence tests. This project organizes activities in the eight areas that are delimited by multiple intelligences: language, mathematics, movement, music, natural sciences, mechanics and construction, social understanding, and visual arts. It is based on the fact that each child has a characteristic profile of multiple intelligences that must be reinforced.

The Key Learning Community Program: Covers both early childhood and high school education and is part of the Indianapolis public school network. At the Key School, several tasks are carried out, which are mandatory: a trade workshop, a visit by a specialist, and student projects. It is based on the premise that the student can build his/her knowledge through discovery learning. Teachers and parents are also involved in this discovery. One of its fundamental objectives is to work daily with the multiple intelligences, in relation to computers, music and kinesthetic-corporal activities. The student must participate every day in a workshop, which allows him/her to master any discipline that is of interest to him/her.

The PIFS (Practical Intelligence for School) Program: Designed for the last years of primary education and the beginning of secondary education. Its objective is to develop practical intelligence, for which it works on understanding and teaching through reflection.

Arts PROPEL (Production, Reflection, Perception, Learning): Focuses especially on working in these three areas: music, visual arts, and creative writing. Its aim was to design a set of assessment instruments to be able to document artistic learning in the last years of teaching in primary and secondary Education.

14.3 Teaching the Use of Multiple Intelligences

A teacher who relies on multiple intelligences to teach mainly carries out these actions:

1. The first step for teachers to make use of multiple intelligences in their teaching activity is for them to identify the intelligences that they have most developed and those that they have least developed. These persons maybe other teaching colleagues or friends, but also the students themselves in order to develop specific activities (Suazo-Díaz, 2006). They can also make use of technological tools that help them in the activities in which they have difficulty.
2. Cognitively challenge students once they have observed them based on the development of different multiple intelligences. This challenge will be carried out by promoting their curiosity, creativity, and initiative. Employs project work methods.
3. Plans the activities, giving enough time for students to work at their own pace.
4. Varies continuously with the method of presentation of the contents used in each case and the different intelligences and combinations between them. The strategies, methods, and educational resources also vary. It provides practical experiences and gets involved and interacts with its students, so that it explores, investigates and discovers with them.
5. It makes clear the rules in an effective way. Uses multimodal systems for learning, using association of content with everyday situations, use of all sensory systems during classes or the execution of expressive activities.

6. It uses resources offered by each type of intelligence to create codes of understanding with its students. For example, a musical stimulus can be used to make way for another activity.
7. Involves technological tools to promote more effective learning environments for the development of all intelligences.
8. It uses the multiple intelligence domain wheel to visualize the relationships between intelligences. The domains that group the different intelligences were proposed by McKenzie (2005) in following way:

The interactive domain is characterized by constant exchange with others and with the environment: therefore, it includes verbal, interpersonal, and kinesthetic intelligences. In this sense, the analytical domain is focused on the analysis of data and knowledge, which includes musical, logical-mathematical, and naturalistic intelligences; and finally the introspective domain, which has a large affective component: intelligences, intrapersonal and visual.

In the construction of the class, procedures, routines, and rules play an important role in providing the class with an order that will allow students to achieve their specific objectives (Armstrong, 2017). The theory of multiple intelligences provides some strategies to control the class. A good example is to see the different ways we can get the class' attention using this theory.

While in the traditional system it would be repeating to the students to be quiet, from the focus of each of the multiple intelligences it would be like this: (a) *linguistic strategy*: offer the message Time to shut up through the devices. You can vary this message to make it more attractive and use witty phrases to awaken students' motivation and attention; (b) *musical strategy*: play music that helps students relax. Examples of this type of music can be different fragments of classical music or a montage with multimedia systems; (c) *kinesthetic strategy*: gestures that evoke silence in the students; (d) spatial strategy: drawing a symbol of silence on devices. (e) *logical-mathematical strategy*: time in seconds that are lost while the students sit down and keep quiet; they can be subtracted for the time of recess; (f) *interpersonal strategy*: use the technique of the broken phone to say in a student's ear: ů"it's time to start, pass it on; (g) *intrapersonal strategy*: start the class and let the students realize that they are behaving badly and keep quiet; and (h) *naturalistic strategy*: make a sound from nature, better from an animal with strident notes.

Evaluation should be taken from an individual point of view that can best detect differences in mental development, learning styles, problem solving styles, as well as evolution in motivation, in the achievement of both personal and sociocultural projects and in the relationship with the environment. Assessment considered from the point of view of multiple intelligences requires detecting and collecting evidence of the modification of the student's behavior.

What they must achieve is to determine the impact that the teacher has had on changing the student's attitudes and behavior, with the ultimate aim that the student acquires the ability to develop metacognition as an instrument that enhances their autonomy. In this sense, evaluation involves assessing the overall performance of the individual by determining what he/she can think or do on his/her own, what he/she needs help with, and what he/she cannot do or think. Evaluation takes place in four stages: initial or diagnostic, formative, summative, and implicit.

14.4 Perception of the Development of Multiple Intelligences

According to a study analyzing the way multiple intelligences are perceived according to gender, there is no appreciable difference between the female and male sexes (García et al., 2018). This study assesses the perception that teachers and families have of the students' abilities in the eight intelligences and compares it with the intelligence results of standardized tests. The standardized tests show that the scores are comparable, with the female students obtaining slightly higher scores. Women are also slightly better assessed on most intelligences by students, teachers, and families.

Students' self-perception: A relevant difference is the way in which women with less visuospatial capacity perceive themselves. There are as many studies that confirm in an objective way this lower capacity in this intelligence of women as there are studies that affirm an equation with men. In the case of the evaluation that teachers and families make of this intelligence in students, the self-perception that students had is not confirmed. In musical intelligence, girls perceive themselves in a higher way than boys.

Teachers' perception: When it is the teachers who value different intelligences, no gender differences are perceived. This result is very favorable and should lead to greater equity, helping to leave behind stereotypes. Such stereotypes maybe behind the lower number of girls than boys in the pool of gifted and talented students. It can be seen that teachers value better in all eight intelligences that part of the student body that performs better, shows greater interest and motivation.

Families' perception: The families' assessment is that they perceive a relationship between the students' abilities and the traditional main intelligences, i.e., linguistic, naturalistic, logical-mathematical, and visuospatial. But they do not perceive a relation between the abilities of the students and the rest of the intelligences. In regard to exceptional abilities, we observe a higher self-perception on an exceptional level in naturalistic and logical- mathematical intelligences in males. On the other hand, women more frequently have a better self-perception in social and musical intelligence. This perception of abilities in the different multiple intelligences of students from three different perspectives provides a great advantage and improves the methods of teaching-learning (Díaz-Posada, Varela-Londoño & Rodríguez-Burgos, 2017).

14.5 Example of the Use of Multiple Intelligences in the Classroom with ICT

Multiple intelligences can help students in the context of ICT use (Figure 14.1). It is even suitable for solving technological problems that can be found in reality (Etchegaray, Guzmán & Duarte, 2017). An example could be established based on a didactic activity focused on achieving the design, manufacture, and assembly of a robot. This also includes programming to control the robot (or Arduino plate).

First session. *First 10 minutes*: the class takes place in the classroom-workshop, so the students must make a change of class, with the attendant noise and distraction. Students can be welcomed with relaxing music such as El Concierto de Aranjuez by Joaquín Rodrigo, thus taking advantage of the opportunity to exercise their musical intelligence. Since the Arduino class is basically practice, it is necessary to form work groups so that each group has an Arduino plate to do the practice. Both in the formation of these working groups and in the subsequent teamwork, we will try to exercise the intra- and interpersonal intelligence. In the small work groups, which will be of three members, it is necessary that all agree on how to carry out the activity, but the next step will be to define a function for each one. For example, while one can type the commands, another can see if there are any faults in the programming sentences and the other can solve any problems with the components and cables, which maybe wrongly pinched on the board, for

Figure 14.1 Use of multiple intelligences in an ICT class.

Source: Cover image.

example. It is important to alternate in the activities so that everyone learns to do all the parts of the practice.

Next 10 minutes: The teacher will make a 10-minute presentation of the theoretical part of the Arduino plate using slides that will be projected. Here you can work on the linguistic, logical-mathematical, and visuospatial intelligences.

Next 10 minutes: To get into the practical part, the Arduino plates and the different electrical components will be distributed among the students. In this process, linguistic intelligence will be taught because the students will identify the concepts by physically seeing the different components. The visual-spatial intelligence will also be exercised because the students will go from seeing the components in two-dimension (2D) to seeing them in 3D.

Next 10 minutes: Once the components and the plates have been distributed, the students will become the protagonists, the explanation will be put into practice, but above all, they will explore and investigate them. First, it is necessary to look for the control program of the Arduino board in the common directory of the students and install it in their computers. Then, we will connect the Arduino board to the computer, exploring the different input ports of the computer and identifying them with the logical ports. This promotes logical-mathematical, linguistic, and visuospatial intelligence.

Last 10 minutes: At the end of each class it is important to collect the material. To do this, the end of the practice is marked and the time to disconnect the boards from the computers, save the programs created and hand the board to the teacher in order with a happy music, which makes the students move and does not slow down this process and the exit of class. The aim here is to emphasize musical and kinesthetic intelligence. Finally, the plaques of each group are collected with the program they have assembled, i.e., without dismantling the components, so that they can continue the practice as they left it, in the next session.

Second session. *First 10 minutes*: The class is resumed by receiving the students with music that helps to calm them down, in this case it could be The four seasons by Vivaldi. Then, the plaques are given to each group.

Next 30 minutes: First a very simple test program is written to check the proper functioning of the components and to familiarize the students with the plate and the different components. In this part, the aim is to prepare the linguistic, logical-mathematical, and intra- and interpersonal intelligence. In the following practice, a circuit with the light-emitting diode (LED) will be assembled so that it can be turned on and off. This implies that a circuit drawn on the blackboard (therefore, it is in 2D) is started and the students have to assemble it on the board, that is to say, pass it to 3D. With this they are exercising the visual-spatial intelligence. In addition, they promote logical-mathematical, linguistic, intra- and interpersonal intelligence. In order to train the students' logical-mathematical intelligence in a more autonomous way, they will have to change the program so that the LED blinks more or less frequently.

Last 10 minutes: Repeated as in the first session.

Third session. *First 20 minutes*: To check that the students distinguish and identify the different components, a taboo game is organized. With this game the students' linguistic intelligence is developed in relation to technological issues. The students are organized in two groups and in each of them they have to choose a member who stands out in the linguistic intelligence. These two students (one from each group) will be in charge of defining to the rest of the group the word indicated by the teacher.

20 final minutes: Another activity is proposed to exercise the kinesthetic intelligence: a circuit with real actors who will be the students is going to be represented. That is to say, one student will take the role of an element that acts as a resistance, another as a switch, another as a power source, another as an LED. And, finally, the leading role will be played by the student who represents the electrical current because he/she is the one who will move through the different components. With this activity the students will represent in a graphic way some abstract concepts such as electric current, voltage drop, and resistance.

Fourth session. In this last session, a circuit is assembled on the Arduino board that works as a thermometer. In this practice, it is used to discover the sensors and, at the same time, to emphasize the naturalistic intelligence. To do this, a plant is shown in class, e.g., an African violet. This plant is an indoor plant and, like any other, needs conditions to maintain itself and grow. It needs a temperature of between 18 and 22°C. Then, once we have built the thermometer with the Arduino plate, we will measure the temperature in several places in the room to see which is the ideal place. In addition, we will take care of it in the following days, checking the temperature every day and choosing the most suitable place. Once the practice has been explained, we will distribute the components among the students so that they can start to assemble them on the board and then program the code.

As this practice can be delayed in time, the next 50-minute session will be used to finish it. At the end of each session the material is collected, including in this process, the disassembly by the students of the components of the plates and they will be given orderly to the teacher for him/her to classify and keep them.

14.6 Conclusions

The theory of multiple intelligences stands out for enhancing students' strengths, skills, and motivation. The students acquire the point of view that they must enhance the intelligences they possess, which are related to the areas in which they are most motivated and, in this way, they can learn what they propose.

The role of the teacher changes significantly with respect to the role he/she adopted in traditional teaching. From being a lecturer of the subject, that is to say, a mere transmitter of knowledge, he/she changes his/her method with the aim of exposing the topics interspersed with the different fields related to the eight intelligences (Sánchez & Llera, 2006). In the application of the theory of the intelligences it is important that the teacher gives the classes in an attractive way and that he/she encourages the students to be the protagonists of the teaching-learning process.

An important part of their task is to help them discover their abilities and then to develop them, i.e., always taking into account the individuality of the students. Studies on Gardner's theory show that its implementation improves the self-esteem of both teachers and students, while detecting a greater increase in interest and participation. This increased interest has an additional advantage of lower absenteeism. But not only are there positive consequences for students and teachers, in the case of families there is also greater involvement (Chen, Isberg & Krechevsky, 2001).

The great contribution of the theory of multiple intelligences is that it helps students achieve knowledge in any field, based on the skills that each one has innate. The theory provides the foundation for this (Eisner, 2004). In the activities proposed in this work, we see an easy way to put them into practice in a secondary education class, regardless of the subject being taught. There are some examples that can be applied to any subject, such as the case of musical intelligence. In contrast, the case of naturalistic or kinesthetic intelligence should be adapted to the context of each subject.

15

Self-Directed Learning and ICT

Celia Corchuelo-Fernández[1], Pilar Moreno-Crespo[2], Aránzazu Cejudo-Cortés[1], and Olga Moreno-Fernández

[1]University of Huelva, Spain
[2]University of Seville, Spain
E-mail: celia.corchuelo@dedu.uhu.es; pmcrespo@us.es; carmen.cejudo@dedu.uhu.es

15.1 Introduction

In the 21st century, with the dizzying and inexhaustible advance of the knowledge society, it is essential to recognize the role that ICTs play in the daily life of any person, to a greater or lesser extent, from buying a public transport ticket to locating the location of a specific corner of the city. Any device with or without an Internet connection opens up a world of social, personal, academic, and professional possibilities. In this century of social networks, it is overwhelming the high number of activities that are no longer necessary to do face to face. We are at the moment of recognizing the technological immersion of our society and using it as yet another resource, of course, also in the classroom.

Self-directed learning and ICTs are linked by the fact that they empower each other. ICTs allow the management of learning by the learner to be offline or online with respect to the physical space offered by the classroom. It allows activities to be carried out individually, interacting between the students or between the student and the teacher. The versatility offered by ICT demands a new organization of the teaching and learning process, as well as that the teacher adapts to a role of guide and companion that encourages self-directed learning and the use of ICT as an effective tool for this.

In this context, one of the methodologies found in growing recognition is the flipped classroom, which allows students to work previously on a content or skill related to the subject under study, reserving space and classroom time

Figure 15.1 Flipped classroom model (Texas Computer Science, 2019).

for activities, tasks, debates, group work, etc., to enhance significant learning and enhance the competence of "learning to learn" required as indispensable for citizenship of the 21st century.

There is consensus among faculty that the flipped classroom is a "learning strategy that offers preparatory or core content outside the classroom and uses class time for active learning" (p.2) (Cobb & Steele, 2014). For teachers, the best way to use classroom time is to support student-centered activities by promoting on/offline interaction (Sung, 2015; Kang, 2015 incorporating cumulative and formative assessment (Forsythe, 2015; Milman, 2012). However, some teachers point out that it could be a fashion because the flipped classroom procedure is based on the premise that the student body carries out preparation tasks before class, which means that there is a high probability of students not being involved with this learning strategy if its programming is not carried out in detail by the teachers (Findlay-Thompson & Mombourquette, 2014).

The flipped classroom model consists of two different moments (Figure 15.1). On the one hand we have a number of tasks that are in the traditional model are usually performed in the classroom. These tasks are reading, viewing audiovisual documents, activities, tutorials, simulations, games, etc. In the flipped classroom model, all these activities are carried out outside the classroom. In other words, the student must perform these activities prior to the classroom session. The time spent in class focuses on activities that enhance meaningful learning. In this way, collaborative projects, individual and group problem solving, and peer-based learning activities can be done. This allows the teaching to be more personalized, with the teacher's decision time being longer for the student body. In this way, the resolution of doubts, orientations, and corrections of the teacher have more attention toward the students than in a traditional teaching model.

The process to plan the teaching following the flipped classroom methodology is divided in two fundamental parts (Figure 15.2). On the

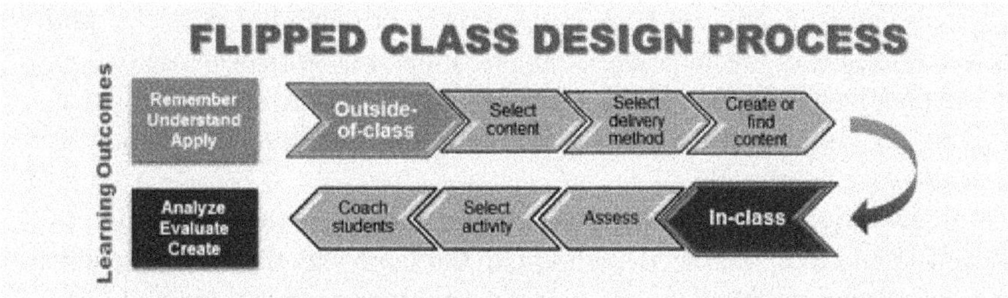

Figure 15.2 Flipped class design process (Texas Computer Science, 2019).

one hand, it is necessary to plan what the students are going to do out of class and, on the other hand, how the knowledge of the material is going to be established as a significant learning in the students (Texas Computer Science, 2019):

- Outside of class: These are the tasks that the student must do previously. These tasks represent the student's first contact with the content of the subject. For this reason the teacher must pay attention to the fact that the medium and format of the content is adapted to the level of the student. This requires the teacher to perform the following tasks:
- Select the content
- Select the form of delivery
- Create or find content
- In class: The teacher must plan the session that will take place in the classroom. To do this, he/she must pay attention to the learning that the student body mainly wants to put into practice. In this way, they can design activities in which the student must seek additional information in real time, research in groups, debate and/or report on what they have learned throughout the process. However, there are three aspects that the teacher should keep in mind at all times:
    - Assess
    - Select activity
    - Coach students

If we go deeper, the results of the research carried out on this flipped classroom methodology are encouraging. It is a methodology that promotes self-directed learning combined with ICT aimed at producing lasting learning. The research focuses on student satisfaction with their own learning process, the improvement of academic results, the potential of collaborative work, the personalization of the teacher's attention to the student, favors interactivity in ICT environments, encourages critical, independent, and reflective thinking in teachers and students, improves student participation, there is a positive perception of the strategies used (they are attractive, active, motivating, dynamic, participatory), and emphasizes autonomous learning, as well as allowing teamwork.

A noteworthy aspect is that the teacher increases the possibility of personalized and individualized attention, in the same way that he/she spends more time solving doubts. Of course, ICTs combined with this methodology and self-directed learning have an impact on motivation and the learning process, facilitating the monitoring of learning and the evaluation process, both for the teacher and for the student.

15.2 Self-Directed Learning

When a learner actively participates in his/her learning by defining his/her own goals and controlling his/her level of effort, motivation, and ability to achieve them, he or she is said to be self-directed in the learning process. In this way, individual freedom, maturity, and responsibility take center stage in the student's learning process (Loyens, Magda & Rikers, 2008; Savin-Baden & Major, 2004); that is, "learning should empower a student to become a free, mature, and authentic self" (p.14) (Savin-Baden & Major, 2004). According to Cerda and Osses (p.1504, 2012), self-directed learning is:

> (...) It is that learning in which the design, conduct and evaluation of a learning effort is decided and carried out by the learner. The key element of this definition relates to the control that a subject has, to decide what to study, how to carry out that task and selecting the type of resources to be used in that process.

In 1978 Guglielmino, in his study on development of the self-directed learning readiness scale, speaks of self-directed learning from two essential approaches: (1) the teacher as mentor or facilitator rather than dispenser of knowledge and (2) the learner as thinker and creator of knowledge rather than a passive receiver. In short, a learner who is capable of learning by himself (Owen, 1999; Paris & Winograd, 2013; Rueda & Dembo, 1995). In this case, two fundamental processes interact: learning and maturation (Vygotski, 1984).

In 2005, Narvaéz and Prada (2015) conducted the research on Self-directed Learning and Academic Performance. The research aims to analyze the results on the readiness for self-directed learning in a group of students from the Peruvian University of Applied Sciences (UPC) and to analyze the level of correlation with academic performance, as well as to explore the motivational factors that might or might not be directing interest toward study in these students. The research concludes that self-directed learning is rather dynamic and evolves according to the age and instructional levels of the subjects. Therefore, the research states that the readiness for self-directed learning responds to a mature factor. This does not exclude that this provision maybe punctual, depending on the course or topic in which the subject is involved.

With respect to self-directed learning and academic performance, we find the work corresponding to 404 undergraduate students from five pedagogical careers (Spanish, mathematics, science, history, and physical education) belonging to a University of the South of Chile (Parra et al., 2014). Asserting that there is an association between self-directed learning skills and academic achievement, they specifically state that students of pedagogy constantly observe their behaviors in relation to the desired level of performance. So students would have the ability to define what, how, and when to monitor their performance and, in this way, take the necessary actions to achieve their academic goals.

This research also provides revealing data in its results. It concludes that educators must know how this ability can be consciously stimulated by them. It identifies techniques, strategies, and principles that allow this type of ability to be promoted from the classroom. Hence the need to generate in teachers the acquisition of pedagogical skills that promote self-directed learning and encourage future graduates to regulate their own learning experiences, since this would be crucial for their personal, social, and work successes.

An example of this is the study carried out by Amador et al. (2007). In this study, it is proposed how to analyze the role of teachers in their role as facilitators of self-directed learning for students of bachelor's degree in nursing from the University of Colima, Mexico. In their results they assure that the tutor facilitates, helps, promotes, or collaborates in the development of the skills that will allow them to be self-directed students in the areas of reflection and critical thinking, information management, and group process, aspects necessary for the good development of self- directed learning.

In the same line, also important is encouraging self-directed learning in the field of health sciences where knowledge is continuously advancing and changing. The importance given to this type of learning is evident, considering it a skill, even seen as a prerequisite for continuous training. It is corroborated by studies such as that carried out by Fasce et al. (2013). Based on the importance given to self-directed learning by educational policies and more specifically by the Association of Medical Schools of Chile as a necessary competence to live in a world of constant transformation, which grants in its discussion a significant correlation between self-directed learning and self-direction, characterized by self-confidence, the search for independence and gratification based on the ability to make decisions. He asserts that subjects who value more directing their own actions also have a greater orientation to learning autonomously, giving self-directed learning the capacity to promote motivated subjects for lifelong learning.

In summary, we can say that self-directed learning adapts to the requirements of the European higher education area, which emphasizes giving the student a leading role in their own learning process. We refer to a protagonist that has the freedom that must be provided by the competence of "learning to learn" and decision-making, with the virtual limits of responsibility over the process and the results of the learning itself, together with the potentials of the growing maturity of each student. We must recognize that the age, level of instruction, and maturity of the student are aspects to bear in mind in self-directed learning. Of course, the figure of the teacher is still fundamental, but in this case as a companion, guide, or mentor who facilitates the environment in which the student must develop his/her learning process (Guglielmino, 1978; Fasce, 2013;).

15.3 Information and Communication Technologies at the Service of the Educational Process

Today, we see how ICTs have become naturalized in our daily uses and customs. It is difficult to be among a group of people among whom there is someone using a technological device. As teachers, we are aware that this is just another resource and that self-directed learning can enhance its functionality through ICT.

In the educational context, ICTs can operationalize the mediation between student and learning; the presentation of content between teacher and student; monitoring, regulation, and control between teacher and student; and the configuration of learning environments for teachers and students (22). Therefore, the potential of self-directed learning is linked to the educational resources that ICTs can provide. It is the logical result of being immersed in the information age, which is growing exponentially. The requirements when it comes to preparing students in skills and knowledge are more demanding and in accordance with the knowledge society and technologies, Driscroll and Vergara (1997, p.84) state that:

> (...) In an information-driven society is the shift from teaching to learning, from teacher-centered education to student-centered or self-regulating instruction. It is a change in epistemology, in what is considered knowledge and how that knowledge is thought to be acquired; it is a change in the metaphor of learning where the student as a receptacle of knowledge becomes a constructor of it.

In relation to the transformations in which Driscroll and Vergara (1997) focuses on teacher training, student self-regulation, epistemology, knowledge, learning, etc., we find the entry of ICT into the learning system, including access to the Internet. Cerda and Saiz (2015), refer that this access to the Internet allows new generations of students to learn in a self-directed way and to carry out training close to individualized learning, in which the presence of the teacher becomes necessary as a companion in this process.

This new role of the teacher, according to Guzmán and Vila (2011, p.3), "makes digital media and open educational resources redraw the boundaries between learners and teachers by weakening the centralization of experience and the distribution of the authority of the subject," in the new knowledge society the teacher is a specialist who is in constant training in parallel on many occasions with his/her own students, recognizing that in that professional growth will never reach an encyclopedic and unique knowledge of the subjects of study, also reinforce new technologies by enhancing the development of skills, creating new spaces of communication, and facilitating the achievement of collaborative learning. These generate different forms of work from those that currently exist in most institutions (Montero, García-Salazar & Rincón-Méndez, 2008).

Proof of these new forms of work is the inclusion of approaches such as flipped classroom, a pedagogical model that aims to reverse the roles of traditional teaching, where the teacher acts as a guide and facilitator of learning, seeking to put the emphasis on multimedia tools and the classroom being the center of development of practical activities, carried out through interactive methods that enhance cooperative work, problem-based learning, and project work (Talbert, 2012). Persky and McLaughlin (2017, p.1) clarifies that "the flipped classroom (also called reverse, inverse, or backwards classroom) is a pedagogical approach in which basic concepts are provided to students for preclass learning so that class time can apply and build upon those basic concepts."

15.4 Research on Flipped Classroom

Works such as that of García-Barrera (2013), *The reverse classroom: changing the response to students' needs*, confirm how this methodology responds to the educational needs of students through the use of ICTs. Among its conclusions, it argues firstly that it favors the approach of didactic sessions that are very different from those we know today. By making the dynamics used encourage students' interest in learning, using technologies that capture their attention and turning them into an active and central part of their own learning process. And secondly, it contributes to the fact that the work of the teacher is not so individual or isolated, by encouraging collaborative work among teachers when implementing sessions, designing materials or exchanging activities, lessons, and educational experiences. It also allows them to devote more time to attending to the diversity of the classroom and to personalize the educational response they offer to each student, adapting it to their individual needs.

This approach is corroborated by studies such as Mendoza-Moreira et al. (2014), designated *Strategies for the implementation of an interactive methodological approach in inverted classrooms for the formation of degree in Education*, whose purpose is to recognize the positive effects of an appropriate methodological approach on the interactivity of students in virtual environments with the collaboration of three groups of undergraduate students in education careers, being the total of participants 91, affirming that the inverted classroom approach favors the interaction between its actors, with equal treatment in the levels of teacher-teacher management. And that the model provides a set of educational strategies that assure the effectiveness of the teaching work, making visible its repercussion in the manifestations of student satisfaction, as well as in their academic results.

Similarly, research such as that carried out by O'Flaherty and Phillips (2015) called *The use of flipped classrooms in higher education: A scoping review*, shows in its results the good performance obtained with the application of this type of approach, so much so that in their conclusions they refer to the fact that it allows teachers to cultivate critical and independent thinking in their students, improves their participation, both inside and outside the classroom, and develops the capacity for lifelong learning, motivating them to obtain good results and preparing them for their future jobs.

Of special interest is the work carried out by García-Rangel and Quijada-Monroy (2015), *The inverted classroom and other strategies with the use of ICTs*, learning experience with teachers, which exposes the differentiated application of learning teaching strategies with ICT support in a community of teachers, who are students of the master's degree in education. One of the objectives of the study was to identify whether the application of the innovative strategy would make a difference in academic results and student satisfaction and whether the academic achievement and satisfaction of the process would be higher in the group in which the inverted classroom was applied. To this end, we worked with two groups, one of them as a control. The results achieved showed greater academic achievement and student satisfaction in terms of the relationship with the teacher and a positive perception of the strategies used to be seen as attractive, dynamic, and motivating for learning.

Along the same lines, we find the pilot inverted classroom experience carried out by Peña et al. (2017), in which it is explained that the inverted classroom model has made it possible to work on two relevant transversal competences; autonomous learning and teamwork, responding to one of the main demands of the students, which is to increase the time dedicated by the teacher to solving problems during face-to-face classes. In addition, it is stated that the fact of focusing the practices using ICT tools, improves the motivation of students and the learning process, facilitating the monitoring of students and the evaluation process and has contributed to a significant increase in the percentage of students who have passed the subject in the first call compared to previous years.

Kim, Park & Joo (2014) investigate the flipped classroom supported by previous study of material prepared by teachers and in the classroom the content is expanded with instant searches in the smart pads, they conclude that this form of work improves the capacity for self-directed learning, the capacity for collaborative learning, and the ability to use information.

The study conducted by Kang (2015) aimed to explore student efficiency and perceptions of flipped classrooms, designed to integrate reading-oriented videos and an activity-centered offline class, for 24 university students. The classroom in which this methodology was not used and those based on the flipped classroom methodology were compared in terms of grammar and vocabulary by means of pre- and post-tests. The results showed that only in the classroom in which the following had been used flipped classroom, there were statistically significant changes in both vocabulary and grammar knowledge. Also that this methodology was positive in terms of satisfaction, help, classroom activities, and teacher roles. However, not doing homework was the biggest obstacle as part of the disadvantages, one of the limitations referred to earlier.

Sánchez-Vera, Solano-Fernández & Gonzalez-Calatayud (2016), present a *Flipped TIC*, a teaching innovation project that was developed at the Faculty of Education of the University of Murcia (Spain). The main objective of this project was the realization of a Flipped Classroom experience in a university classroom. The students had to perform production tasks (consult doubts, debates, creation of resources, realization of practices, etc.). The results showed that the satisfaction of the students with this type of methodology was very good, although they still considered totally necessary the figure of the teacher in the classroom. The experience concluded that the use of this type of methodology should be promoted in the classroom to encourage significant learning among students.

Galway et al. (2014) carried out the flipped classroom methodology, analyzing the experience and comparing the results with those of previous years. The study concludes that the student deepens the knowledge, being a positive experience about the learning and perceptions about the methodology carried out. Although the data reveal that the result in the evaluations was similar to that of previous years, where traditional methodologies were used, the students' evaluations of their experience in the course, their learning, and the increase in self-perceived learning were better.

15.5 Conclusion

We emphasize that the presence of ICT in classrooms must be normalized, as the blackboard did in its day as an element present in a pedagogical model. There is no doubt that ICTs offer us resources that have to adjust to the needs of teaching and submit to methodologies that facilitate the promotion of a model of the citizen of the future, whose qualities make him/her capable of "learning to learn," capable of developing his/her autonomy and personal initiative, as well as maintaining a critical perspective, among other issues. To do this, it is essential to have methodologies that focus on promoting the role of the student in his/her own learning process, giving him/her autonomy and responsibility over it.

Faced with this situation, we have highlighted a working methodology that unifies self-directed learning and ICT. As we have already discussed, it is the flipped classroom. This methodology is very versatile and can be adapted to a wide variety of contexts.

Thus, after the various studies we have analyzed, we can say that the flipped classroom based on self-directed learning and supported by ICT:

- Enhances the role of the student in his/her own learning process
- Promotes autonomy, collaborative work, and participation.
- Encourages interactivity in digital environments.
- Improves student satisfaction and motivation.
- Improves academic results.
- Enhances responsibility and decision-making through self-directed learning.
- The teacher personalizes the attention and increases the time for the resolution of doubts.
- Encourages critical thinking in teacher and student.

In short, we conclude that flipped classroom is a methodology that is changing the traditional teaching-learning model. It achieves collaborative learning environments, encourages student interest in learning, and encourages the use of new technologies. It is a didactic strategy that ensures a change in the way the teacher works; by encouraging cooperative work between teachers, reinforcing personalized and individualized attention in the classroom. It also guarantees generating better academic results and hence guaranteeing future graduates greater adaptation to future work contexts.

16

Videogames in the Teaching of Social Sciences

Emilio José Delgado-Algarra

University of Huelva, Spain
E-mail: emilio.delgado@ddcc.uhu.es

16.1 Introduction

Videogames can show relevant problems and social issues through the playful development of attractive historical representations for students; establishing five categories related to the teaching and learning of social sciences (Cuenca & Martín, 2010):

- Games of economic nature
- Games of social nature
- Games of geographical nature
- Games of artistic nature
- Games of historical nature

Similarly, continuing with the aforementioned authors, some of the social sciences contents that can be worked with videogames are related to the following:

- War and conflicts
- Urban and territorial management
- Democracy and citizenship
- Economy and trade
- Environment

The technological development is expanding the immersion in playable experiences, in addition to the didactic possibilities of video games with the support of different peripherals and applications. Both Virtual reality

Figure 16.1 Some applications/videogames integrated with VR/AR technologies.

(VR) and augmented reality (AR) are examples of new ways of immersion (Figure 16.1).

In terms of VR, we have devices such as HTC or Oculus Rift, we also have mobile applications. Despite its possibilities and the relative ease of using VR in educational environments, in its most immersive, it requires peripherals to recognize the movement of the body that needs an important economic investment. This technology still to this day is in development and presents a significant room for improvement.

In terms of AR, we have devices such as Microsoft Hololens and numerous mobile applications and videogames integrated with that technology. AR is an emerging technosocial technology that has demonstrated its practical effectiveness in both university and nonuniversity educational contexts (Cabero et al., 2017).

In general terms, given the technological development (AR, VR, three-dimensional (3D) printers, mobile applications, digital platforms, etc.) and the growing interest of students in videogames, numerous educational researchers have highlighted both the advantages of gamification as the responsible inclusion of videogames in educational contexts (Figure 16.1).

This chapter focuses on videogames without didactic objectives useful in learning environments, finishing with a review about the possibilities of game series Civilization and SimCity for the teaching and learning of social sciences, history, geography, and civics.

16.2 Benefits of Gamification and Videogames in Teaching and Learning Environments

In recent years, the use of video games has been shown in different age ranges. In this sense, as shown in the report "The New Faces of Gaming" of the Interactive Software Federation of Europe (ISFE) (2017) through the GameTrack survey (ISFE and Ipsos Connect), a survey that has covered the European markets of the United Kingdom, Germany, France, and Spain, it is confirmed that approximately three quarters of young people between 6 and 24 years old say they have played video games. On the other hand, and continuing with the same report, it should be noted that the popularization of the use of smartphones and tablets has contributed significantly to expanding the age field of players in recent years.

In this sense, in relation to gamification, it:

> *seeks to stimulate participation and involvement in an activity through the stimulus derived from the challenge linked to obtaining achievements and satisfaction related to the receipt of rewards throughout the game. For this reason, it has been progressively incorporated into the classrooms, trying to stimulate learning. The dynamics will work best when it allows to establish a progression so that each time a challenge is overcome, a new one is proposed. The progression of challenges and the establishment of the constant incentive system constitute two major difficulties for the design of gamified activities in the classroom* (2017: 5, translated by Delgado-Algarra).

Some videogames without explicitly educational purposes can be used as a resource for the teaching and learning of social sciences and history in different educational stages. For example:

- Games such as Libertus help us understand the Roman world (Delgado-Algarra, 2018).
- Sagas such as SimCity allow us to work in an integrated management of space and resources (Delgado-Algarra, 2014, 2019).
- Games like This War of Mine allows us to work on raising awareness about issues related to memory (Delgado-Algarra, 2014).

Definitely, gamification responds to a process of implementation of playful procedures in educational environments (Ayén, 2017) and advantages of using video games in teaching have been reflected in classroom experiences (Lorca Marín & Vázquez-Bernal, 2012).

In a broad sense, games have levels of benefits for learning as a main reference for the didactic review of videogames in block 3 as listed hereunder (Martínez-Navarro, 2017):

- Support to the internalization process of multidisciplinary knowledge. For didactic review of videogames about: [knowledge].
- Understanding of ways of thinking different from those of our environment. For didactic review of videogames: [positioning].
- Improvement of the abilities to solve problems. For didactic review of videogames: [resolution].
- Improvement of strategic planning. For didactic review of videogames: [planning].
- Encourages decision-making. For didactic review of videogames: [decision].
- Develops social skills, experience with various identities and experiences. For didactic review of videogames: [social].
- Improves attention and concentration. For didactic review of videogames: [attention].
- Increases motivation. For didactic review of videogames: [motivation].
- Improves critical thinking. For didactic review of videogames: [reflection].

16.3 Videogames as Didactic Resources for the Teaching of Social Sciences

Regarding the introduction of video games in the classroom as a teaching resource, some general strategies are the following (Heick, 2018) (Figure 16.2):

- Play at school: The selection of the game, preferably, is decided between the teacher and the students, justifying the reason for the choice.
- Students play at home: rejecting the use of video games involves a total disconnection with the environment of students who, for the most part, are accustomed to the use of games or the consumption of mobile applications.
- See them playing: related to the previous proposal, since not all students have access to video games outside the classroom or do not have adapted equipment, the possibility can be raised that the recorded gameplay will be edited and uploaded to Youtube by adapting privacy.

- Analyze: there is the possibility of looking for information about the games, their developers, their development, or their tendencies; being able to analyze video reviews, text, post on social networks, related art, or movements linked to games.
- Reimagine: as a novel, comic, short, etc., what would change the game, justify the reason for the changes, evaluate the impact of those changes, and how they would benefit the learning of concrete contents of social sciences, geography, history, and civics.
- Plan: plan videogames requires considerable effort and time, in this case, the integrated approach with other areas would be advisable.
- Create: there is the possibility that students create their games, with programs such as Scratch, or that they can create their environments and games, with, e.g., Minecraft edu.
- Facing relevant social problems are treated in some videogames and may have a positive impact on the development of social and citizen commitment and the enrichment of critical consciousness. Such problems include the following:
  - civil consequences of war (This War of Mine),
  - resource management (SimCity Edu),
  - understanding the motivations that determine the decision-making in political-military strategies (Nobunaga Ambition, Age of Empires, Civilization, etc.), etc.

We are going to review the didactic possibilities of two sagas of great trajectory in the world of videogames:

- Sid Meier's Civilization/Civilization: began its journey in the market in 1991, and since then, on the market, there are a total of six historical titles and one about the future. Civilization is a turn-based strategy game in which we manage an empire, choosing from a wide variety: Spanish, Japanese, Romans, Carthaginians, Egyptians, Indians, Incas, Greeks, Japanese, Koreans, Chinese, Mongols, Ottomans, Persians, Russians, Sioux Indians, Zulus, Vikings, Portuguese, Sumerians ... among many others. Each empire has a real leader and aims to be the leading civilization in the world.
- SimCity: SimCity is a city management simulator that became very popular in the 1990s that gives us control over different variables for the design and management of the city, always adapting to the budget and taking into account the action of possible natural disasters. In general,

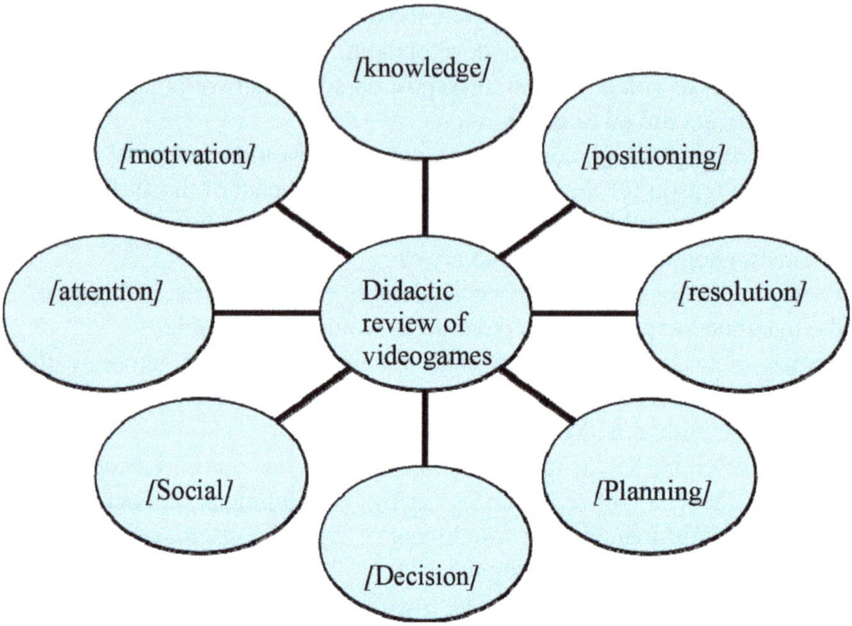

Figure 16.2 General categories for didactic review of videogames.

SimCity allows the user, depending on their level of deepening in the game, to get involved in their experience with different levels of difficulty.

These games involve simulations, decision-making, management, resource management, and artificial intelligence, a strategy where comprehension, planning, execution, and revision processes come into play. From this point, start the didactic review of videogames, where we will discuss the following categories and questions (Figure 16.2):

- [Knowledge] Are there useful concepts for social science learning? Which are?
- [Positioning] Faced with certain events in the game, can we adopt a positioning? Does our positioning influence the playable experience? Can we understand the thinking of others?
- [Resolution] Are presented relevant socioenvironmental problems
- Are competencies set in motion to solve problems?
- [Planning] Are there planning phases for the action? Does it allow us to evaluate advantages and disadvantages before putting it into practice?

- [Decision] Is decision-making encouraged? On what decisions are made? What are the difficulties in decision-making?
- [Social] What is the role of social and civic competence? Are issues related to identity raised? How are the social events that can be experienced in video games?
- [Attention] Must we focus our attention? What must we focus on?
- [Motivation] What is the motivation to continue playing? How can this motivation improve social sciences learning?

Didactic review of videogames about [Knowledge]

Civilization raises the development of real civilizations and real events with relevant historical figures. It includes issues that allows to work the historical time from a holistic perspective such as (Figure 16.3):

- Technological research.
- Influence of religion
- Political relations
- Diplomatic relations
- Military relations
- Economic relations

As for SimCity, the management of the economic and urban heritage requires a process of economic optimization with multiple variables, among others:

- Expenditures for the execution and maintenance of infrastructures and services
- Income from taxes (with decision-making associated with the rise or fall of them)
- Tolls and use of services
- Sale of hydroelectric resources

Likewise, it requires management of the amount of population; soil value; and mortgage loans with interest; management of the territory and spaces creating residential, commercial, and industrial areas.

16.3.1 Positioning

Both Civilization and SimCity allow to understand different motivations and different points of view, being able to serve as a starting point to raise debates about relevant socioenvironmental problems in the classroom.

The player must be involved in the game experience and must understand the variables involved in historical events, in the case of Civilization, and

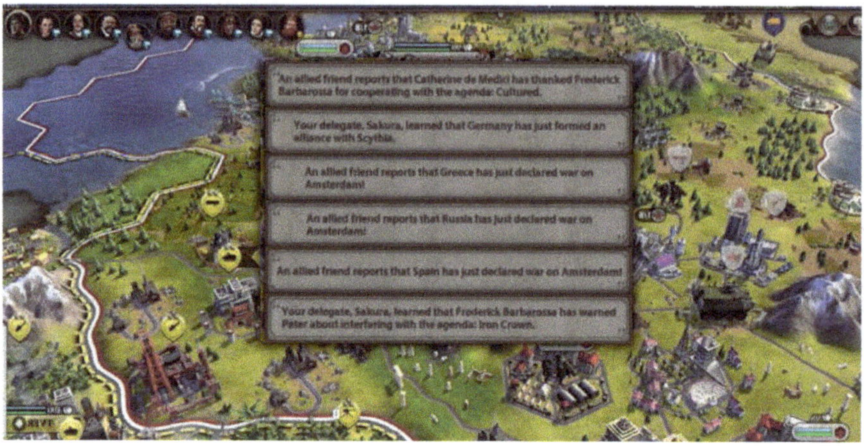

Figure 16.3 Diplomacy and alliances in Civilization VI (2016).

in the management of the city that is created, in the case of SimCity, at the same time that explores and identifies the relationships between them (Figure 16.4).

16.3.2 Resolution

In both cases, the player is exposed to problems that, in many cases, are a consequence of their decision-making. The student must make responsible use of the tools and means at his/her disposal in the game to solve the problems that arise.

The use of this videogame in the classroom allows to contextualize and apply in the virtual practice the knowledge acquired by the students previously, including:

- Development of attitudes of critical thinking
- Development of attitudes of problem solving

Regarding SimCity, with the first roads and buildings, problems will begin. We will have little money and we will have to cover the essential needs for the city. Covered some essential needs for the city to work, there are many tasks to be done, such as providing basic services such as health, safety, education, or firefighters (Figure 16.4).

Figure 16.4 Example of city focus on culture in SimCity (2013).

16.3.3 Planning

Closely related to the above, in both cases, attending to the problems requires prior planning, evaluating pros and cons of our decisions, and assuming the consequences.

Regarding SimCity, while getting involved in the gaming experience, the player must understand the number of variables involved in the management of the city he/she is creating, while exploring and identifying the relationships between them (examples: soil quality, public services, cost of maintenance of services, and amount collected).

16.3.4 Decision

In Civilization VI, decisions are made in multiple sections, being especially relevant those related to (Figure 16.5):

- Technological research
- Influence of religion
- Political relations
- Diplomatic relations
- Military relations

Figure 16.5 Research and technology decisions in Civilization VI (2016).

- Economic relations
- Many other variables

In relation to SimCity, as the player's degree of understanding about all the variables and their interrelations grows, this one goes from making decisions without a defined orientation to planning the next steps to take at two levels:

- Microlevel (e.g., organization of new small areas or services)
- Macrolevel (e.g., change of type of energy source, optimization of the city's water supply system)

For which it must establish a strategy of organization of expenses and planning of the cost of maintenance of the facilities in front of the obtained collections. Space is very limited so as to cover the deficit in something—safety, health, energy, tourism, etc.—we must obtain it from neighboring cities.

16.3.5 Social

In both cases, situations arise in which the social relations and the decisions taken in this regard have a decisive influence, either in the historical development, in the first case, either in the city development, in the second.

The social and civic competence is very important in Civilization and SimCity, but Civilization also includes historical environments. On the other hand, SimCity allows the user, delving into the game, to get involved in their experience with different levels of difficulty, where space is limited and services must be negotiated with other cities that offer resources that we lack.

16.3.6 Attention

Connected with the overcoming of the problems that arise and the observation of the results obtained as better decisions are made.

In the execution phase, in Civilization and SimCity, the player triangulates the possibility of implementing his/her decisions and experiences the possible unforeseen consequences of his/her decisions.

Regarding SimCity, there are six major specializations in which it is convenient to focus on advanced parties:

- Culture, creating great monuments, sports stadiums or exhibition centers, which attract many tourists
- Mining, extracting coal and minerals, which we can consume and sell to other cities
- Drilling to extract oil
- Games, to build casinos that attract many tourists, but also delinquency
- Electronics, to be a neuralgic center of high technology and development of processors, computers and televisions to later export them
- Trade, to import, export, and store resources

For example, if you decide to build an electronic city, it is necessary to prioritize education and create a nearby university community as soon as possible (Figure 16.6).

16.3.7 Motivation

In both cases, the integration of different variables and decision-making generate consequences that can be analyzed within the classroom, either through classroom play or through the recording and observation of games. The fact of being able to see the evolution of the respective civilizations or of the city, respectively, facilitates the meaningful analysis of the causality and a deeper understanding of the role of the different variables proposed in the game to extrapolate them to the reality in view of the teaching and learning of history and social sciences.

Figure 16.6 Creation of a university community in SimCity (2013).

16.4 Conclusion

Although gamification is not a new concept, we consider that the technological development and the recognition of its educational possibilities have extended the possibilities in a new way, contributing significantly to education, in general, and to the teaching and learning of social sciences, in particular. The development of technologies and the expansion of the use of tablets and mobiles in the daily context of students should be considered in education.

It could be considered that addressing this reality of the student, beyond encouraging motivation and contributing to the understanding of social sciences contents, serves as guidance for responsible and critical use of them, taking advantage of their educational possibilities.

In addition, the succession of challenges to overcome that arise in videogames seeks to enhance participation and involvement in the proposed tasks, aspects that we consider especially relevant for the teaching of social sciences, geography, and history, areas where a commitment to relevant socioenvironmental issues is required.

According to the reviews of the series Civilization and SimCity, great possibilities are observed for the teaching and learning of history, geography, and social sciences, attending to concrete contents of the area, to the resolution of problems, planning, decision-making, development of social skills, attention, motivation, and reflection.

17

Gamified Learning Environments

Diego Vergara, Rodríguez[1], José María Mezquita Mezquita[2], and Ana Isabel Gómez Vallecillo[1]

[1]Santa Teresa de Jesús Catholic University of Ávila, Spain
[2]Santa Teresa de Jesús Catholic University of Ávila, Spain, and IES Maestro Haedo, Spain
E-mail: diego.vergara@ucavila.es; unmezquita@gmail.com anai.gomez@ucavila.es

17.1 Introduction

Currently, one of the major fields of study in education is being monopolized by educational gamification. Although the term was coined in 2003 (Pelling, 2015) and was rapidly diffused into different sectors, marketing (Zichermann & Linder, 2010; Werbach & Hunter, 2012), tourism (Xu, Weber & Buhalis, 2014), health (King et al., 2013), etc., it is precisely in recent years that the greatest amount of papers is being published on the topic in education journal articles (Yusoff et al., 2018). The logged results in Figure 17.1 show a sustained increase in the number of articles dealing with the issue of gamification over the last decade.

Within the field of education, the benefits accrued from the synergy of ICTs and the large numbers of virtual tools (VTs) that have appeared in recent years, help create a gamified learning environment (GLE). Among these VTs are those related to multiple-choice questions (MCQs), e.g., Kahoot, Quizizz, Plickers, Socrative, Poll Everywhere, Quizlet, QuizUp, QuickKey, Edulastic, Wooclap, Quizalize, Go Pollock, etc., which currently represent one of the directions in which education is moving, given the attraction of young people to the virtual world. In fact, some of these tools, defined as *game-based*

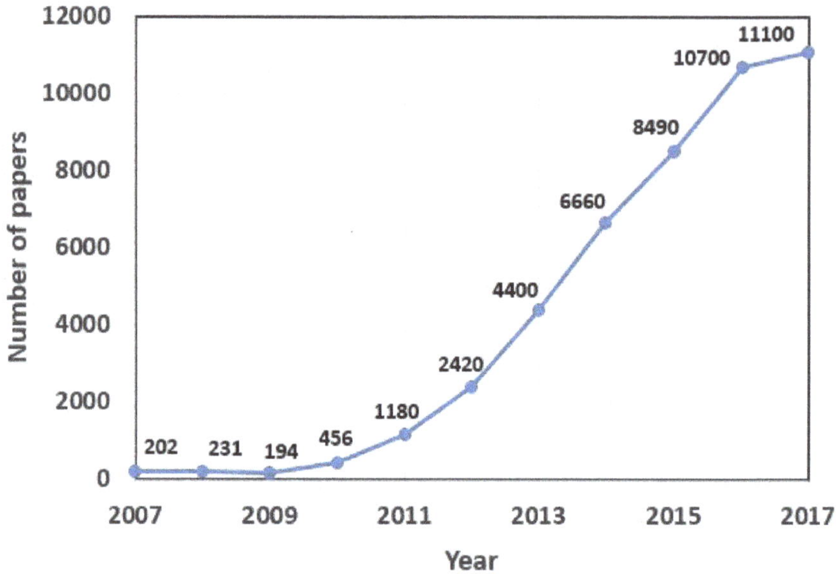

Figure 17.1 Number of papers indexed in Google Scholar regarding gamification (data collected in March 2018).

student response systems (GSRSs) in previous studies (Wang, 2015), are designed specifically for mobile devices (Bicen & Kocakoyun, 2017).

This chapter addresses the importance that GSRSs are acquiring in the teaching-learning process and analyzes its main characteristics, examining some of the tools that seem to arouse more interest in the education sector: Kahoot!, Quizizz, Plickers, and Socrative. Furthermore, this study corroborates that this type of VTs meets the necessary requirements to create a GLE using a suitable methodology designed for this purpose. Finally, the data collected in this chapter show that an *educational trend* is developing based on the use of the MCQs VTs that helps build a GLE.

17.2 VTs Generating a GLE

GSRSs are used at all levels of education, from the preuniversity level (Premarathne, 2017; Pede, 2017) to the university level (Solmaz & Çetin, 2017; Sánchez-Martín & Dávila-Acedo, 2017; Cheong Cheong & Filippou, 2013). Among all the VTs designed to pose MCQs (in a gamified environment), some of the most widely used and recognized in education are

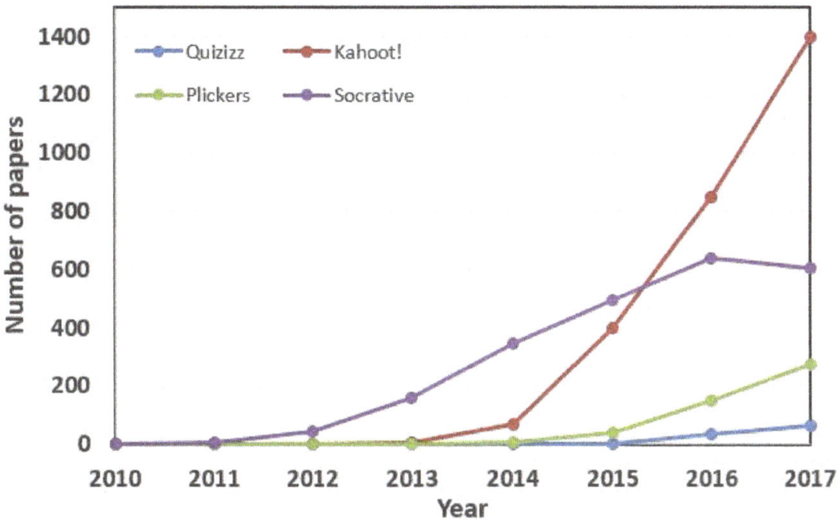

Figure 17.2 Number of papers indexed in Google Scholar regarding the selected four VTs (data collected in March 2018).

Kahoot!, Quizizz, Plickers, and Socrative (Plump & Özdamlı, 2017; Krause, O'Neil & Dauenhauer, 2017; Bello & Merino, 2017). The selection of these four GSRSs has been made following two main criteria: (i) the diffusion and intensive use of them in education and (ii) the speed and ease of use in the classroom. As in the case of educational gamification (Figure 17.1), Figure 17.2 shows the growing trend in the use of this type of VTs.

17.2.1 Kahoot! (https://kahoot.com/welcomeback/)

One of the most used GSRSs based on MCQs is Kahoot! The origin of Kahoot! is dated 2013, when entrepreneurs Johan Brand, Jamie Brooker, and Morten Versvik carried out a joint project with the Norwegian University of Technology and Science (Kahoot, 2018).

They were later associated with Professor Alf Inge Wang, and later joined by the Norwegian entrepreneur Åsmund Furuseth. In March 2013 Kahoot! was released in a private beta version on SXSWedu, making it one of the most preferred mobile application for university students (Bicen & Kocakoyun, 2017).

Kahoot! is a web-based service that, together with its application for mobile devices, promotes educational processes based on gamification; in

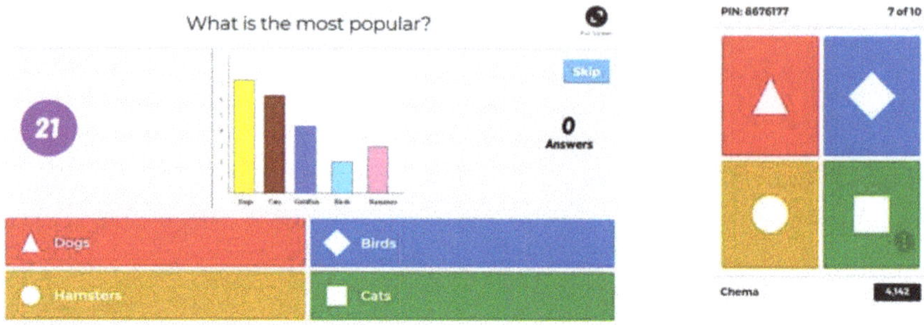

(a) (b)

Figure 17.3 Print screen of Kahoot!: (a) projector; (b) smartphone or tablet.

other words, it works as a competitive knowledge game. The platform includes a web-based creator tool that let educators create MCQs and ordering questions (Figure 17.3). The questions are grouped into questionnaires and can be shared on the community page or through Facebook, Twitter, Pinterest, Google+, or e-mail. Educators can, in turn, search for and select public quizzes created by others.

The implementation of Kahoot! in the classroom is simple, the teacher only needs a projector and a digital device (computer, tablet, or smartphone). Students use their own devices to press the same color and symbol as the answer displayed on the screen to indicate the correct answer to the given question. To access each questionnaire, students must enter the Kahoot! App or open the website https://kahoot.it/ and enter a game pin. This application requires a coordinated rhythm in answering the questions, as they are designed for all students at the same time. Once the questionnaire has been completed, the system generates a scoreboard of the students based on the right answers they have given. Likewise, the answers given by each of the students can be collected as an Excel document.

The main advantages of using Kahoot! in the classroom are the following: (i) the speed and ease in the use of the application and (ii) the existence of a large bank of questions and questionnaires accessible to any user. The main drawback is the possibility of students copying each other's answers, as all questions are projected simultaneously.

17.2.2 Quizizz (https://quizizz.com/)

Quizizz is a web-based tool created by Ankit Gupta and Deepak Joy Cheenath in February 2015 (Quizizz, 2015). It does not require any special equipment

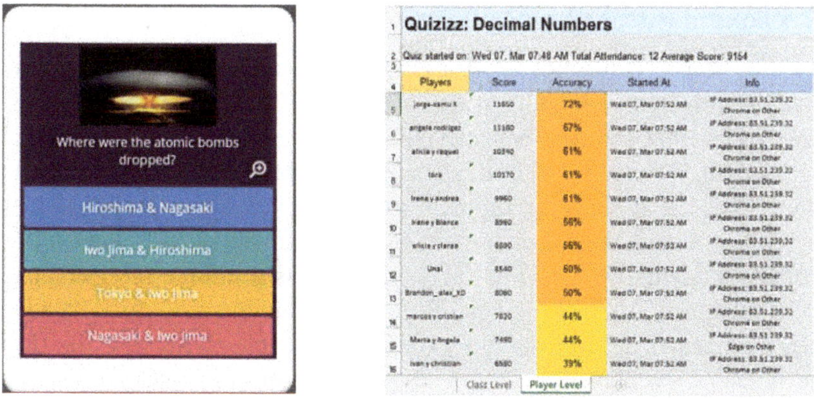

(a) (b)

Figure 17.4 Print screen of Quizizz: (a) smartphone or tablet; (b) Quizizz report.

to be used, just one device with a web browser for each student (tablet, smartphone, PC, or laptop) and another for the teacher. Quizizz is accessible via mobile devices or via its website. To implement it in the classroom, the teacher must create the quiz he/she would use to test their students' knowledge in the application or on the Quizizz website or import a quiz from an Excel document. The software allows for MCQs with one or more correct answers (Figure 17.4a). To carry out the activity with Quizizz, students must enter the website https://quizizz.com/join/ and insert the code of the questionnaire provided by the teacher and then their names. The fact that Quizizz allows questions and answers to be presented in a random order for each participant means that each student can work at his/her own speed. Since students are unlikely to be answering the same question at the same time, this makes it difficult for them to copy each other's answers. The teacher monitors the progress of each student at all times. Once the activity is finished, the tool presents a leaderboard showing the three best players and a report is sent to the teacher, both in the application and in a spreadsheet (Figure 17.4b), which includes detailed information of students' responses and the number of correct answers given for each question.

The main advantages of using Quizizz are the following: (i) fast and easy access to be implemented in the classroom; (ii) possibility of randomized questions, thus making it difficult for students to cheat; (iii) existence of a question bank accessible to the user; (iv) possibility of integration with other learning management systems and formats (Edmodo, Google Classroom, Remind, and Excel); and (v) quality of the reports provided by the software.

17.2.3 Plickers (https://www.plickers.com)

Although the prototype was developed in 2008 by Nolan Amy when he was a math teacher in Richmond, CA, it was actually in 2013, under the tutelage of ImagineK12, that Plickers became publicly accessible (Plickers, 2018). Among the four VTs analyzed in this chapter, Plickers is the least technical demanding tool, as it simply requires a mobile device and a projector for the teacher and a card printed on a sheet of paper containing a unique visual code for the students (Figure 17.5a). Each card has a unique number that should be assigned to each student. Before starting the activity with the students, the teacher designs the questionnaire (multiple-choice test) on the website https://www.plickers.com/, or through the app for mobile devices. The teacher puts the questions up on the projector from his/her mobile device. Students use their cards that can be turned in four orientations, each of which gives a different response. To answer the question, students have to select one of four possible answers—A, B, C, or D—by turning the orientation of their card. When all students hold up their cards with their choice of answers, then the teacher proceeds to collect all the data by scanning them using the phone's camera (Figure 17.5b). Instantly, a report is displayed with information that indicates correct and incorrect responses. As in the case of Kahoot! all students are asked each question at the same time. Once the whole questionnaire is completed, the teacher receives a file in Plickers mobile app

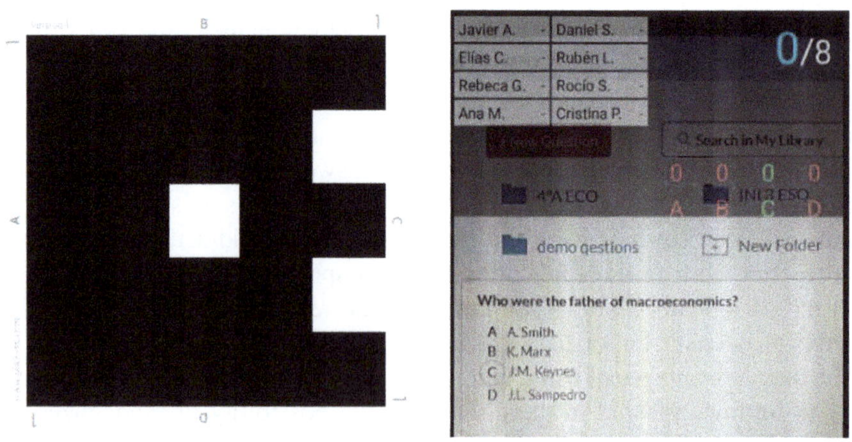

(a) (b)

Figure 17.5 Print screen of Plickers: (a) Plickers card; (b) teacher's smartphone or tablet.

that groups all the answers given by the students and the scoreboard of the participants.

Plickers is particularly suitable for schools with limited technical resources and for young students in primary and first years of secondary school, where it is forbidden to bring a smartphone to class, i.e., no form of bring your own device (BYOD) strategy is implemented in the classroom. As a disadvantage, identical to the case of using Kahoot!, the fact that all students answer the same questions at the same time may encourage them to cheat.

17.2.4 Socrative (https://www.socrative.com/)

Socrative was created in 2010 by Professor Amit Maimon, from the School of Management and Massachusetts Institute of Technology (MIT) (Bello & Merino, 2017). This VT has a separate application for teachers and students: Socrative Teachers and Socrative Students, respectively. The technical requirements for a session with Socrative are similar to those of Quizizz, only one mobile device is needed for each participant. For its implementation in the classroom, the teacher must have previously completed the questionnaire on the Socrative website or in the teachers' app. The types of questions that the teacher can elaborate are multiple choice, true or false, and short answer questions. Once the questionnaire has been designed, the teacher submits it for the students to take individually (Quiz) or in teams (Space Race). To access the questionnaire, students must log in to Socrative Students and enter their teacher's room code. The teacher manages student input and the functioning of the game through his/her interface (Figure 17.6a). Students can see both the question and the answer options on the device (Figure 17.6b). At the end of the activity, the teacher receives a detailed report with the classroom's overall score, graded responses, and class score per question, together with a scoreboard of the participants.

The main advantages of using Socrative in the classroom are the following: (i) availability of different question options; (ii) possibility for students to complete the questionnaires in one go or independently; (iii) capability to import questions from a spreadsheet; and (iv) possibility to include feedback after each question. However, compared to other GSRS, the process of importing other teachers' quizzes is more complicated.

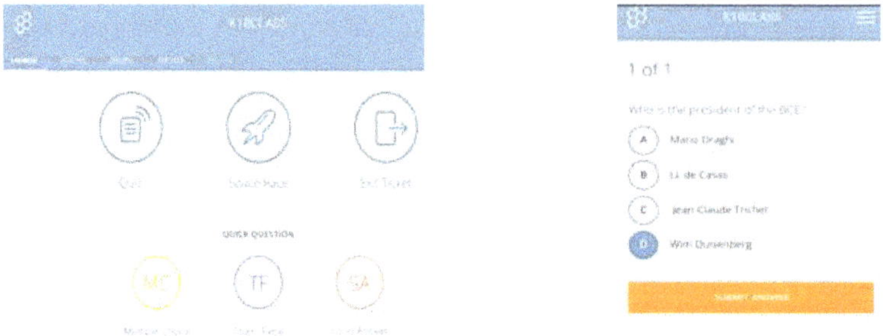

(a) (b)

Figure 17.6 Print screen of Socrative: (a) teacher interface; (b) student's smartphone or tablet.

17.3 Implementation of Gamification in Education

The widespread implementation of gamification in education (Figure 17.1) is triggering the emergence of different digital resources that contribute to the generation of a GLE. In addition to showing a clear demand from the education sector, some entrepreneurs have seen a business opportunity with the premium or paid versions apps.

A number of factors highlights the increasing interest arouse by GSRSs based on MCQs (favoring a GLE) in education, among them the following are important: (i) the proliferation and mass use of GSRS in the last 10 years (Figure 17.7), as shown in variables such as number of users, number of quizzes uploaded, and number of apps downloaded (Table 17.1) and (ii) its educational effectiveness, supported by numerous studies and a wide acceptance among students.

Regarding the number of apps downloaded, Table 17.1 shows the public data available on the Google Play Store (data in Apple's App Store is not included as it is not public). Although the values shown in Table 17.1 are a clear indication of the use of this type of VTs, most of these GSRSs can be run directly on the internet (e.g., Plickers, Quizizz, Kahoot, etc.). The lack of access to App Store data and the direct use of VTs on the Internet suggest that the data reflected in Table 17.1, despite being high, represent a lower average than the real one.

Other indicators suggesting the educational success of a GRSR are the number of users and the number of public quizzes uploaded, etc. In this regard, Kahoot! has over 70 million active monthly users and 1,6 billion

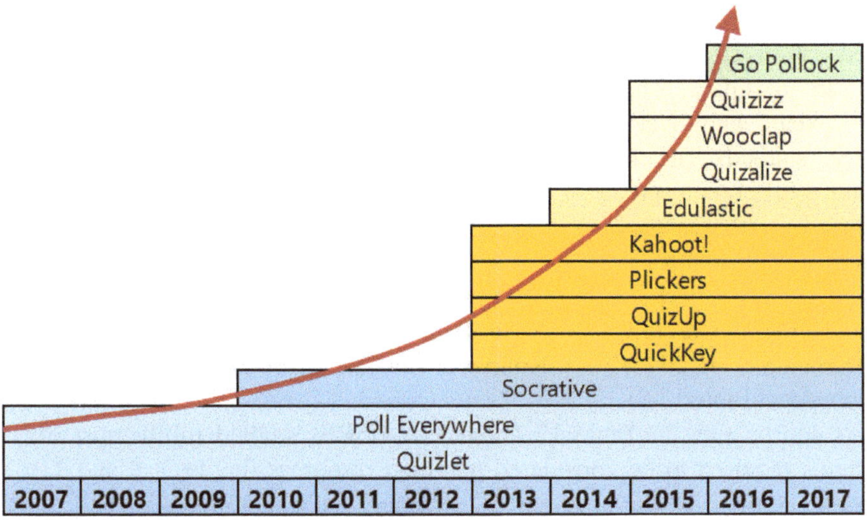

Figure 17.7 Proliferation of GSRSs (VTs based on MCQs).

Table 17.1 Google Play downloads of several GSRSs.

GSRS	Google Play downloads
Kahoot!	10.000.000
Quizlet	10.000.000
QuizUp	10.000.000
Socrative Student	1.000.000
Socrative Teacher	100.000
Plickers	100.000
Quizizz Students	100.000
Poll Everywhere	100.000
QuickKey	10.000
Edulastic	No app
Wooclap	No app
Quizalize	No app
Go Pollock	No app

players since it was launched (data collected in April 2018) (Kahoot, 2018); Plickers presents 338,302,672 student responses, 19,305,047 questions asked, 1,089,965 registered teachers, and a dissemination of quizzes from 190 countries (data collected in April 2018) (Plickers, 2018). Socrative is now being used by about 1.1 million teachers and millions of students worldwide (Matheson, 2018).

Due to the increasing application of gamification-based methodologies in education (Figure 17.1), some efforts are being made to integrate VTs in a GLE. This chapter presents an analysis to assess to what extent gamification elements are being introduced in the four GSRSs studied. The list of game elements devised by Werbach and Hunter (2012) is used as a starting point to analyze and discuss the requirements that a tool must fulfill to be considered gamified. To this purpose, a literature review addressing these requirements has been carried out (Werbach, & Hunter, 2012; Sobocinski, 2018; De-Marcos, García-López & García-Cabot, 2016; Rocha Seixas, Gomes & de Melo Filho, 2016; Barata et al., 2016; Tsay, Kofinas & Luo, 2018), obtaining a total of 19 components (Table 17.2).

As can be seen in Table 17.2, the four GSRSs studied fulfill most of the items required to be considered gamified resources (between 9 and 11 out of 19 items). All the analyzed VTs comply with approximately half of the required items. Taking into account the great similarities that the whole group of GSRSs exhibit, these findings can be extrapolated to the rest of tools. In addition, even though this type of tools does not directly comply with some of these components (data with an asterisk in Table 17.2), many of them can be compensated by using a coherent and motivating narrative, with a certain degree of freedom for the students (Werbach & Hunter, 2012; Sobocinski, 2018). This would imply that the percentage of compliance with the requirements to be a gamified VT would increase even more, approaching 90% for all GSRSs. All this indicates that there is a distinct interest in this type of VTs being designed with game elements, thereby showing an educational trend toward the use of gamification in the classroom.

Evidence of this educational trend can be seen in a series of factors that have taken place over the last decade. On the one hand, we observe an increase in the use of gamification-related methodologies (Figure 17.1). On the other, the recent appearance of VTs including game elements (Goshevski, Veljanoska & Hatziapostolou, 2017), especially those based on MCQs (Figure 17.7). The latter VTs have a number of common features such as the following: (i) they are newly created tools, most of which have been launched in the last five years (Figure 17.7); (ii) they have been rapidly disseminated and accepted by the public, as shown by the app download data (Table 17.1), registered user data and quizzes uploaded; and (iii) they are designed taking into account the game components, as stated in Table 17.2.

However, it should be noted that the use of VTs by itself does not generate a GLE, but they must be used within a teaching methodology that

Table 17.2 Requirements for a VT to be considered as gamified (3,24-28). (*The lack of this component can be compensated with an appropriate narrative.).

Components	Description	Kahoot!	Quizizz	Plickers	Socrative
Achievements	Defined objectives	Yes	Yes	Yes	Yes
Avatars	Visual representations of a players' character	No*	Yes	No*	No*
Badges	Visual representations of achievements	No*	No*	No*	No*
Boss Fights	Especially hard challenges at the culmination of a level	Yes	Yes	Yes	Yes
Collections	Sets of items or badges to accumulate	No*	No*	No*	No*
Combat	A defined battle, typically short-lived	Yes	Yes	Yes	Yes
Content Unlocking	Aspects available only when players reach objectives	No	No	No	No
Gifting	Opportunities to share resources with others	No*	No*	No*	No*
Leaderboards	Visual displays of player progression and achievement	Yes	Yes	Yes	Yes
Levels	Defined steps in player progression	No*	No*	No*	No*
Points	Numerical representations of game progression	Yes	Yes	Yes	Yes
Quests	Predefined challenges with objectives and rewards	No*	No*	No*	Yes
Social Graphs	Representation of players' social network within the game.	No	No	No	No
Teams	Defined groups of players working together for a common goal	Yes	No*	No*	Yes
Virtual Goods	Game assets with perceived or real-money value	No*	No*	No*	No*
InstantFeedback	Information about how the player is doing.	Yes	Yes	Yes	Yes
Freeware	Available for free	Yes	Yes	Yes	Yes
Customizable	Element customization	Yes	Yes	Yes	Yes
Easy management	User-friendly management	Yes	Yes	Yes	Yes

takes great care of the narrative. Furthermore, gamification based teaching practices have a different impact on student achievement, since the success of this methodology is conditioned by both student personality (Buckley & Doyle, 2017) and student learning style (Sobocinski, 2018; Buckley & Doyle, 2017). The use of gamification tools in the classroom is such a recent phenomenon that more research is needed, because although most of the empirical studies confirm the positive effects on students learning outcomes (Dicheva et al., 2015; De-Marcos, Garcia-Lopez & Garcia-Cabot, 2016; Yildirim, 2017), other authors challenge the common belief of the benefits obtained with games (Judson & Sawada, 2002; Domínguez et al., 2013). Thus, it is estimated that in the coming years the evolution of studies related to educational gamification will maintain the same growing trend shown in Figures 17.1 and 17.2.

17.4 Conclusions

There is a growing tendency toward the implementation of methodologies based on educational gamification. Because of this, many of the current educational virtual resources are designed to include game components, thus favoring the possibility of generating GLEs. This chapter reveals the emergence of a *new educational trend* that encompasses the educational use of VTs based on MCQs (Kahoot!, Quizizz, Plickers, Socrative, Poll Everywhere, Quizlet, QuizUp, QuickKey, Edulastic, Wooclap, Quizalize, Go Pollock, etc.). Although this type of VTs are designed with a high percentage of gamification components, it cannot be assured that these resources promote educational gamification by themselves, since what really favors the process of gamification in the classroom is the methodology used.

18

Virtual and Augmented Reality

José Gómez Galán[1], Eloy López Meneses[2], César Bernal Bravo[3], and Esteban Vázquez-Cano[4]

[1]University of Extremadura, Spain, and Ana G. Méndez University, Puerto Rico-USA
[2]Pablo de Olavide University, Spain
[3]King Juan Carlos University, Spain
[4]UNED, Spain
E-mail: jgomez@unex.es; jogomez@uagm.edu; elopmen@upo.es; cesar.bernal@urjc.es; evazquez@edu.uned.es

18.1 Introduction: VR and AR for ICT Integration

We live in a society where daily interactions are increasingly conditioned by ICTs, and where learning is conceived as a cocreation of knowledge in technology-enhanced communities (Kali, Baram-Tsabari & Schejter, 2019). In this sense, we live in an era of immense advances in ICTs that affect all aspects of our lives, and especially what we know and how we learn (Hoadley and Kali, 2019). Consequently, to the above, the technological imperative in education is related to a general evolution and digitalization of society and to the need for new skills (Engen, 2019).

Today, the pedagogical use of ICTs has facilitated the shared creation of knowledge through learning communities (Romero & Patiño, 2018) and a set of benefits and potentialities with respect to traditional methods of content transmission (López-Belmonte et al., 2019). Similarly, the socioeducational and technological changes of the 21st century have contributed to promoting profound transformations in higher education institutions aimed at strengthening new trends that seek to favor the mobility of students, graduates, and teaching staff, among other aspects, to give way to a competitive society based on knowledge and new technological trends, where students become fundamental actors in achieving this purpose (Veytia, Gómez-Galán & Morales, 2019). In turn, as Delgado-Vázquez et al. (2019) point out, these play a role of enormous wealth and education professionals cannot remain on the sidelines.

In the same vein, knowledge and skills with the use of ICTs are an essential part of modern life (Juhaňák et al., 2019; Matosas-López et al., 2019). Furthermore, these technologies have the potential to prepare students for life in the 21st century. Through learning these new skills, students are prepared to face future challenges that must be based on a proper understanding of their world. The use of ICTs can help students develop the skills needed for today's globalization. This is because they can help students to develop their skills, increase their motivation, and expand their knowledge and information (Grabe & Grabe, 2007; Hussain, Morgan & Al-Jumeily, 2011). Therefore, one of the main approaches to the educational process is to focus on learning and the student from an educational approach based on digital skills, with the conviction that it will contribute to improving university training processes. Similarly, it can be inferred that the curricula leading to the obtaining of a degree should, therefore, have among their objectives the acquisition of them by the student body, as well as the procedures for evaluating their acquisition.

To speak of *competencies* is to mention a polysemic concept characterized not only by the diversity of semantic meanings that have been attributed to it over time, but also by the uses that have been made of this concept in different training scenarios and its link with other terms such as strategy, skill, etc., which make the idea itself diffuse (Cebrián & Junyent, 2015).

Different authors indicate that a competence is a process in which people can creatively solve problems, carry out activities, formulate questions, search for relevant information, analyze, understand, and reflect when applying their knowledge by giving an answer to the demands of a real environment (Serrano, Biedermann & Santolaya, 2016; Ramos, Chiva & Gómez, 2017). At the same time, digital competence can be understood as the training of knowing how to use technology effectively to improve all areas of our daily lives.

However, digital competence is not an isolated skill to be developed, but rather it is a compendium of skills, abilities, and attitudes before different areas and dimensions of knowledge (Rodríguez-García, Raso-Sánchez & Ruiz-Palmero (2019), where the protagonist of the educational action is the student body, who in turn, must face up to this entire technological society and which has transformed the different ways of communicating, learning, accessing work, etc., in short, living the present and being prepared for the future (Gisbert and Lázaro, 2015).

In this context, both VR and AR can be extremely important in the integration of ICTs into educational processes. VR is an interactive simulation based on a computer-created environment where one or more users can enter virtually, using different tools that allow them to experience multisensorial perceptions, emulating reality, and creating a state of presence in the environment. There are four basic elements that make up VR that involve both the virtual system (hardware) and the user: virtual environment, sensory feedback, interactivity, and virtual presence (Sherman & Craig, 2003). Thus, a virtual environment is a reconstruction of the real world through the use of 3D modeling software and rendering engines for the application of physical laws that are applied to give a greater sense of realism (Bystrom, Barfield & Hendrix, 1999).

AR, on the other hand, and within the concept of VR, refers to the direct or indirect visualzsation of elements of the real world combined (or augmented) with virtual elements generated by a computer, whose fusion gives rise to a mixed reality (Cobo & Moravec, 2011). In the same line, Azuma (1997) conceives it as a technology that combines real and virtual elements, creating interactive scenarios, in real time and recorded in 3D. It is also defined by Cabero et al. (2016) as that environment in which the integration of the virtual and the real takes place, i.e., the combination of digital information and physical information in real time through different technological devices; i.e., it consists of using a set of technological devices that add virtual information to the physical information, thus creating a new reality, but in which both real and virtual information play a significant role in the construction of a new mixed, amplified, and enriched noncommunicative environment. In addition, AR optimizes learning processes and increases the interest and participation of students (Cawood & Fiala, 2008) interacting with elements in different dimensions and enhancing the use of emerging technologies in educational ecosystems.

18.2 VR in Education

Because of this, it is essential that from educational contexts, a study of the didactic possibilities of VR begins to be developed (Figure 18.1). There are currently many lines of research that are focusing on studying the effects that the new information and communications society is producing in different educational contexts and curricula, but few of them focus on

VR (Fabris et al., 2019; Kamińska et al., 2019; Vesisenaho et al., 2019; Huttar & Brintzenhofeszoc, 2020). It is undeniable that the recent prominence of computer technologies forces, and of course, their integration into the school. Although there is a broad consensus on this, the different opinions are reflected in the question of how this can be done, i.e., how these new and VTs (which are used today in practically all human technological dimensions) of so many implications should be used in the present and, of course, in the future (precisely the time for which current school children are being trained) within the field of education (Gómez Galán, 2001, 2002, 2014). This means that VR must soon and seriously be considered, although it is not yet fully extended if it is to be so in a short period of time.

Logically, the educational possibilities offered by VR are enormous (Liu et al., 2017a; Kang & Kang, 2019; Ahir et al., 2020; Radianti et al., 2020). Because it is a rapidly developing and spreading technology (the limit is only set by the amount of computer and telematic resources it needs, but the power of computers is continuously growing so that its great moment will soon come) we do not yet know what its limits are, although we now have sufficiently consistent examples. In the field of education, it is now present in various university and professional environments (where there is equipment prepared for this), generally to develop simulations of various activities (in civil aviation, architecture, etc.) that are economically more profitable than their development in reality (we also hope that VR can be very productive from an ethical perspective, e.g., for experiences in the field of medicine or surgery, avoiding having to perform practices on human beings or animals).

Likewise, the relevance that it may have in society, being used by different media, especially by video games, makes it necessary for it to be known by teaching professionals (Gómez Galán, 2014; Burkle & Magee, 2017; Adarve et al., 2019). It is clear that today we can not only consider these technologies as resources or teaching aids, but it is necessary to take them into account as important elements and facets of our society, which must be known, critically analyzed, and understood as any other aspect (within politics, economy, culture, etc.) of the present civilization, in the effort to form people with criteria of discernment in a complex and sometimes chaotic current world (Gómez Galán, 2002). In particular, we consider that it is necessary to open up currents of educational research that will allow us not only to establish practical strategies to strengthen the presence of ICTs in the teaching-learning contexts, but also to make them as simple as possible (starting from the undeniable fact that there is still a lack of teacher training—both in primary and secondary education—in this field) to adjust to the authentic needs of

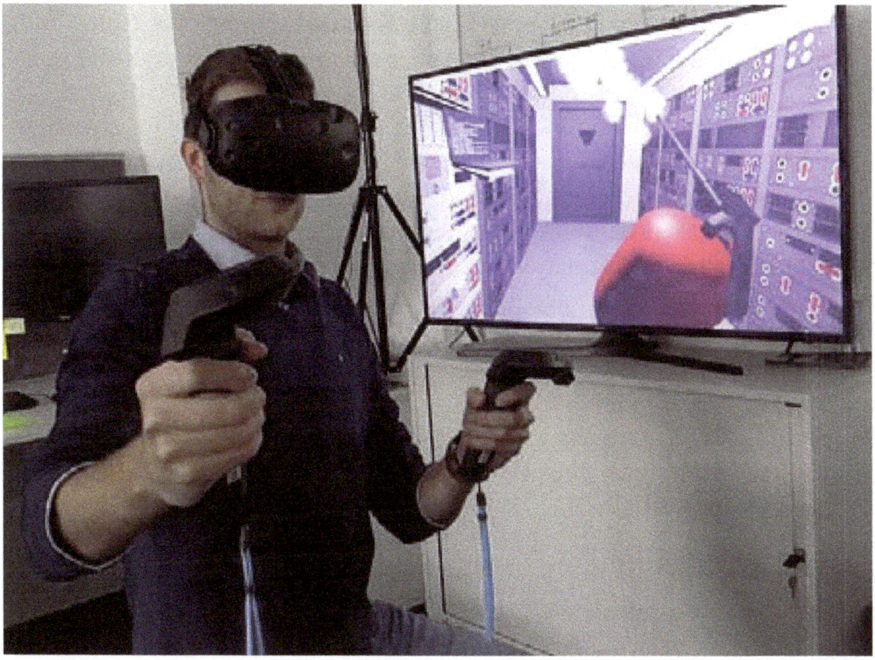

Figure 18.1 A researcher using VR headset to investigate ideas for controlling rovers on a planet.

Source: European Space Agency http://www.esa.int/spaceinimages/Images/2017/07/Reality_check

today, all within a framework that can offer us the greatest educational advantages. From this point of view, and due to the implications that VR may soon have in the lives of children and young people, it is essential that educators know about it and analyze it in such a way that they can create critical attitudes that favor their development at this age, in these children, and help them to overcome the reality-fiction differences, whose limits are beginning to be confused even in the most traditional media (Gómez Galán, 2001, 2014).

18.3 Didactic Possibilities of AR

Within the set of VR and what it brings in didactic scenarios, what AR can offer right now more educational possibilities in education (Gómez Galán, 2014; Hood, 2017; Moreno-Martínez & Leiva, 2017; Fernández-Robles,

2018), especially in higher education (Chang et al, 2013; Barroso & Gallego-Pérez, 2017) and is growing in recent years thanks to mobile digital devices that make it accessible to the entire public (Cabero and Barroso, 2018).

Different authors (Johnson et al., 2016; Cabero et al., 2019) express that AR is the environment in which the integration of the virtual and the real takes place, i.e., the combination of digital information and physical information in real time through different technological devices. This technology also allows information to be displayed by providing multimedia materials or texts linked to objects or places, in a simple and immediate way (Billinghurst, Kato & Poupyrev, 2001) and to address all human senses of perception, however, the most widespread variation of AR is commonly the representation of visual virtual information added to the real environment (Cheng & Tsai, 2012; Maquilón, Mirete & Avilés, 2017). AR offers numerous educational possibilities and immense potential for enhancing learning and teaching (Dunleavy, Dede & Mitchell, 2009; Bacca et al., 2014; Prendes, 2015). In addition, it provides users with access to rich, varied, and meaningful multimedia content, giving them a relevant context in which to interact immediately (Han et al., 2015; Cabero & Garcia, 2016) (Figure 18.2).

Figure 18.2 Chromville AR.

Source: Imascono. https://vimeo.com/166031138

For example, the AR helps students to conduct real-world explorations. By displaying virtual items next to real objects, it facilitates the observation of events that cannot be easily observed with the naked eye (Wu et al., 2013). In this way, it increases students' motivation and helps them acquire better research skills (Sotiriou and Bogner, 2008). According to Dunleavy, Dede & Mitchell (2009), the most significant advantage of AR is its unique ability to create hybrid and immersive learning environments that combine digital and physical objects, thus facilitating the development of processing skills such as critical thinking, problem solving, and communication through interdependent collaborative exercises. And it helps to activate cognitive learning processes, the development of cognitive and spatial skills in students, regardless of their age and academic level, as well as favoring more motivating, collaborative, and interactive learning scenarios (Cabero et al., 2016). Finally, several authors (Fernández -Robles, 2018; Cabero & Garcia 2016) indicate among their limitations the lack of teacher training and the scarcity of resources and learning objectives focused on AR.

Recent publications such as those by Bacca et al. (2014), Gómez Galán (2014), Billinghurst et al. (2015), Chen et al. (2017), Dunleavy and Dede (2014), Palmarini et al. (2018), and Wu et al. (2013) have shown that AR research encompasses a broad spectrum of objectives and methods. In turn, the United States, United Kingdom , and China Taipei are the countries with the largest number of scientific publications that are addressing the study of VR in education (Liu et al. 2017a,b).

On the other hand, several studies revealed that the use of AR in formal education could enable educators to combine these applications with various pedagogical approaches, such as situated learning (Chang & Jen-ch'iang, 2013; Crandall et al., 2015), research-based learning (Wang et al., 2014; Bressler & Bodzin, 2013; Chang, Wu & Hsu, 2013), and play-based learning (Hwang et al., 2015; Lu & Liu, 2015). It is used today at all levels of education, from K-12 (Chiang, Yang & Hwang, 2014; Kerawalla, Luckin, Seljeflot & Woolard, 2006) to university level (Ferrer-Torregrosa et al., 2015).

18.4 VR and AR Learning Scenarios

In general, VR and AR encourage a proactive learning environment (Fombona, Pascual & Madeira (2012)) and lead to high student satisfaction (Han et al., 2015; Kim, Hwang & Zo, 2016). Similarly, the use of AR is useful in the construction of emerging competencies on the use of ICT,

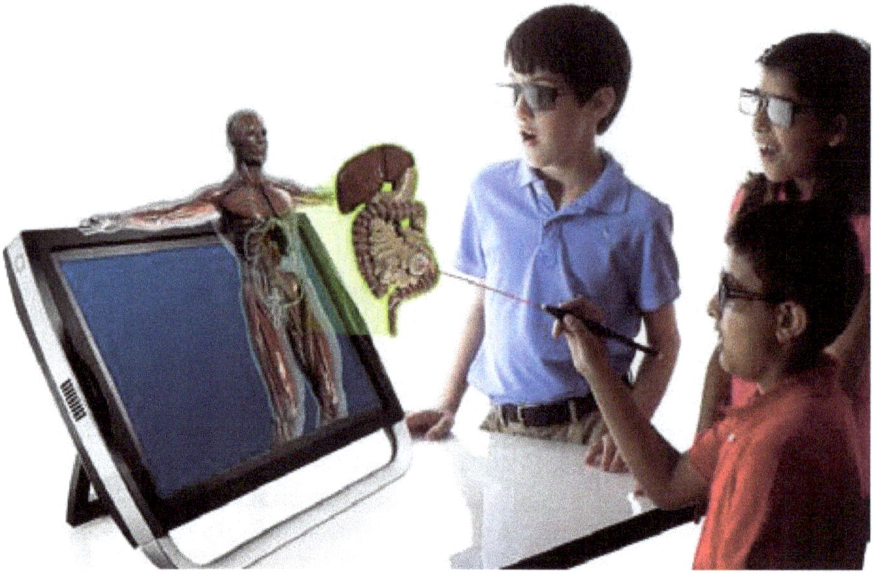

Figure 18.3 zSpace All-in-One for Education. CC BY-SA 4.0.

Source: https://en.wikipedia.org/wiki/ZSpace_(company)#/media/File:Anatomy_kgroup.jpg

teamwork skills, the discovery of new immersive teaching resources useful in new educational scenarios, unknown by most of the students, which can help the development of new training processes from an investigative, constructivist, and ubiquitous perspective (Cabero, Vázquez-Cano & López-Meneses, 2018).

According to Fernández -Robles (2018), AR can be presented as a truly useful technology for university training, since it allows working with active methodologies that offer the possibility of visualizing the object from different perspectives, facilitates the acquisition of knowledge that is difficult to access, enables the presentation of simulated scenarios, and enriches printed material. Likewise, the perception of the students of higher education in the areas of social sciences has shown in their appreciations that the use of activities based on the AR makes possible a greater motivation and reflection and increases the positive attitude to learn the contents (Chang et al., 2013).

The use of VR and AR is useful in building emerging ICT skills, teamwork capabilities, and the discovery of new useful immersive learning resources (Figure 18.3). VR- and AR-based learning activities can be particularly useful in preuniversity and higher education studies related to

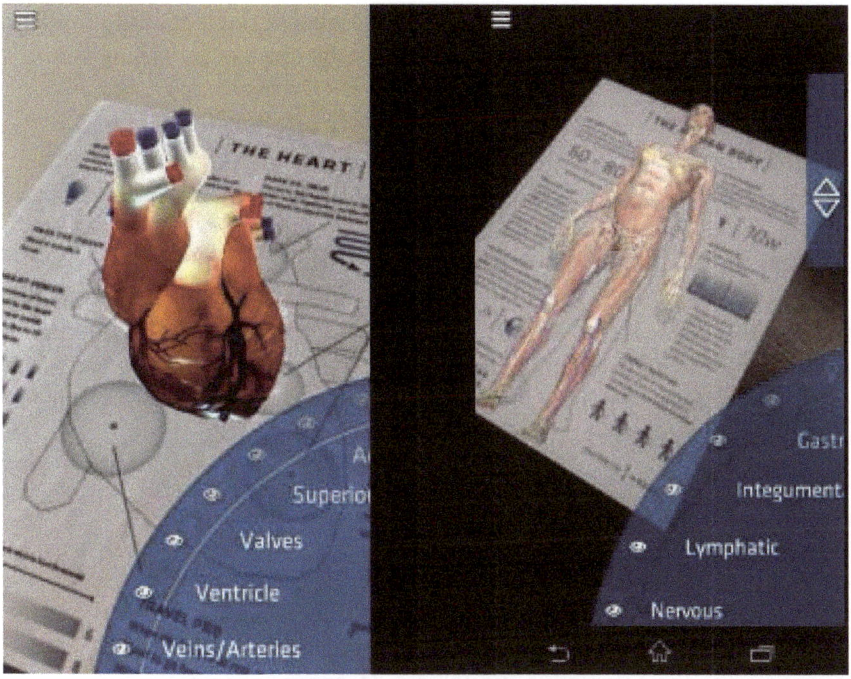

Figure 18.4 Anatomy4D AR.

Source: Citecmat. http://citecmat.blogspot.com/2014/12/anatomy-4d.html

the humanities and social sciences areas, as it allows access to content, often presented in a more unidirectional way through printed media that does not increase motivation or require a greater effort of abstraction among students. With this resource, the contents are visualized in a more creative, dynamic, and real way of achieving a more attractive, innovative, and motivating didactics at any educational level.

Despite these supposed benefits, we believe that the proposals and models for integrating VR and AR should be further developed, taking into account that their development should be properly contextualized: taking into account the issues and resources available and what the teacher training is, as without the appropriate training, their use can be counterproductive and can lead to a delay in the development of skills and content in the teaching program (Gómez Galán, 2001). Moreover, we must bear in mind the resources available to educational institutions and students themselves in order not to generate a greater digital gap or discrimination for economic reasons or for access to technology (Figure 18.4). Likewise, it is necessary to guarantee that this technology incorporates mechanisms of usability and adaptability to the different special needs that students may have in the classrooms (Hung et al., 2013).

It also allows the observation of experiments or phenomena that occur after a long period of time (months, years, decades, etc.) in seconds, such as, e.g., Mendel's laws, although it also allows the opposite case by facilitating the observation of what happens in an instant (López-Belmonte et al., 2019). In this sense, AR is presented as a truly useful technology for university education, since it allows working with active and constructivist methodologies, offers the possibility of visualizing the object from different perspectives, facilitates the acquisition of knowledge that is difficult to access, makes the presentation of simulated scenarios possible, and enriches printed material (Fernández-Robles, 2018).

18.5 Conclusions

Although many advantages of both VR and AR have been related, not everything is positive in the use of these tools. Some limitations of both are the following: (a) virtual sociability should be encouraged, as it is now less humane; (b) the interactive digital divide can be increased; (c) more training is needed for educators in the didactic use of these systems; (d) very few VR and AR systems are adapted to the special needs of the students; and (d) finally, the implementation of emerging VR and AR technology in educational institutions can have a high financial cost.

But they should certainly be integrated into education. Carrying out training experiences of this kind helps to expand the theoretical didactic model known as Technology, Pedagogy, Content and Space (TPeCS) (Kali et al., 2019). TPeCS argues that good teaching today requires not only "an understanding of how technology relates to pedagogy and content" (as in Mishra and Kohler's TPaCK, 2006, p.1026), but also an understanding and ability to adapt existing physical spaces, take advantage of alternative spaces, or design new ones.

This may include the use of digital technologies such as mobile devices, VR and RA, large digital screens, etc. (Tissenbaum and Slotta, 2019). It is therefore time to begin the transformation from digital literacy to digital competence. Its acquisition must be done through initial training practices and promote new training scenarios where the learner becomes a "prosumer" of technological messages.

In short, it can be considered that the challenge for education, within this framework, lies in redesigning its training matrices around professional skills rather than around traditional subjects in such a way as to promote the development of didactic proposals that involve collaborative work for the promotion of significant learning (López-Meneses et al., 2019) and the progressive increase in the use of electronic educational resources in teaching (Vázquez-Cano, López Meneses & Sánchez-Serrano, 2015; Infante-Moro et al., 2017). Ultimately, VR and AR can become an interesting and efficient technosocial emerging trend in educational contexts and help to generate processes of curricular innovation in the classrooms of the 21st century.

19

Information Resources in Education and Educational Research: A Brief Guide

Ángel M. Delgado-Vázquez

Pablo de Olavide University, Spain
E-mail: adelvaz@bib.upo.es

19.1 Introduction

The ease of creation and dissemination of digital information has brought with it the birth or conversion of a multitude of sources and information resources that are available to be used in the fields of educational research and innovation. Bibliographic databases, digital libraries and repositories, primary data banks, but also educational materials: from exercises to complete courses, through didactic units, evaluations, readings, and all kinds of resources. And all this, to a large extent, is freely available to teachers, students, and researchers in the educational field.

This volume of resources—an undeniable source of knowledge—often contrasts with the difficulty that educational agents and students experience in covering the different typologies of materials available and the main access to them. Therefore, this chapter establishes a classification of available basic resources (Table 19.1) organized in bibliographic resources, data sets, and educational materials. In addition, notable examples of each type of resource are incorporated with the aim of facilitating a useful guide—not exhaustive—indicative of the possibilities that open for the generation of knowledge in the area of education and educational research.

Presented hereunder are each of the resources, whose description has been made based on the section "*About us*" or similar to each one of them, checking and introducing additional information when it has been considered pertinent.

Table 19.1 Basic Resources and Access

Resource type	Subtype	Resource	Access link
Bibliographic resources	Bibliographic Database	**ERIC**	https://eric.ed.gov/
		Education Research Complete	(https://www.ebsco.com/products/research-databases/education-research-complete)
		British Education Index	https://www.ebsco.com/products/research-databases/education-research-complete
		IRESIE	http://iresie.unam.mx
		REDINED	http://redined.mecd.gob.es
		HEDBIB	https://hedbib.iau-aiu.net/home.php
	Digital libraries	**OECD i- library**	http://www.oecd-ilibrary.org/education
		IBEDOCS	http://www.ibe.unesco.org/
	Education Plans and Policies Databases	**Planipolis**	http://planipolis.iiep.unesco.org/
		Eurydice	https://eacea.ec.europa.eu/national-policies/eurydice/home_en
	Open Access Repositories	**BASE**	https://www.base-search.net/
		CORE	https://core.ac.uk/
Data sets	DataRepositories	**UIS.stat**	http://data.uis.unesco.org/Index.aspx
		EdStats	http://datatopics.worldbank.org/education/home
		OECD Online Education Database	https://www.oecd.org/education/database.htm
		National Center for Education Statistics	https://nces.ed.gov/
Educational materials	Open Educational Resources Repositories	**OER Commons**	www.oercommons.org
		Merlot	www.merlot.org

Source: Self-elaboration.

19.2 Bibliographic Databases

The information resources included in this section are the online bibliographic databases, the evolution of the old summary bulletins which, in paper format and later in digital format, were created and distributed by documentation centers and specialized libraries in the field of research and educational management (Table 19.1).

19.2.1 ERIC (https://eric.ed.gov/)

This is a bibliographic database created and maintained by the Educational Resources Information Center of the US Department of Education in the United States. It stores reviews, journal articles, meetings, official documents, theses, reports, audiovisuals, bibliographies, directories, books, and monographs specialized in sciences of the education since 1966. ERIC is dedicated to scholars, researchers, educators, policy-makers, and the general public and provides full access to much of the documentation related to education in the interest of improving research, and consequently, educational practice. Currently it contains more than 1.7 million references, of which more than 1 million are peer-reviewed, while more than 400,000 are full text accessible for free.

19.2.2 Education Research Complete (https://www.ebsco.com/products/research-databases/education-research-complete)

This is a bibliographic database produced and maintained by the EBSCO Information Services covering scholarly research and information relating to all areas of education. Topics covered include all levels of education from early childhood to higher education, and all educational specialties, such as multilingual education, health education, and testing.

Education Research Complete also covers areas of curriculum instruction as well as administration, policy, funding, and related social issues.

The database provides indexing and abstracts for more than 2,300 journals, as well as full texts for nearly 1,400 journals. This database also includes full texts for almost 550 books and monographs, and full texts for numerous education-related conference papers.

19.2.3 British Education Index (BEI) (https://www.ebsco.com/products/research-databases/british-education-index)

Created and maintained by the BEI office at Leeds University between 1986 and 2013 and taken over by the EBSCO in 2013 covering articles dating back to 1950, the British Education Index includes over 270,000 articles published in UK journals from 1929 onwards, and over 11,000 records for UK theses.

BEI covers areas such as cognitive development, computer-assisted learning, curriculum, educational policy, educational psychology, educational technology, management in education, multicultural education, physical education, science education, special educational needs, and teacher education.

19.2.4 IRESIE (http://iresie.unam.mx)

The Índice de Revistas de Educación Superior e Investigación Educativa (*IRESIE*) is an information system specialized in Ibero-American education. Created in 1979 at the National Autonomous University of Mexico (UNAM), it aims to identify, organize, and disseminate the results of research, teaching or planning that are generated in the field of education in the Ibero-American region. The database contains more than 90,000 bibliographic records of research reports, theoretical or review articles, essays, institutional reports, biographies, interviews, statistics, book reviews, and other educational documents published in more than a thousand titles of specialized journals in education and complementary disciplines published in Mexico, Latin America, the Caribbean, Spain, and Portugal.

19.2.5 REDINED (http://redined.mecd.gob.es)

This is an educational information system created in 1985 by the Spanish Ministry of Education through the Centre for Educational Research and Documentation, which contains several databases integrated into a single repository.

As a bibliographic database it includes the delivering of the articles published in the Spanish education journals of greater impact and visibility, as well as others that universities or teacher training centers publish. There is also a section that includes doctoral theses, and master and degree theses. Other documentary types, such as monographs, conference proceedings, and reports, are also described.

As for educational resources, REDINED serves as a repository, but also as a collector of other Spanish educational repositories in which memories of teaching innovation projects, guides, manuals, and all kinds of materials for teaching are deposited.

19.2.6 HEDBIB (https://hedbib.iau-aiu.net/home.php)

International Bibliographic Database on Higher Education (*HEDIB*) is a unique resource of references and publications on higher education systems, administration, planning, and policy and evaluation from around the world. Links to electronic publications are provided where freely available. New higher education publications are identified by the IAU Information Centre monitoring team which has been managed by the International Association of Universities (IAU) since 1988, and receives contributions from UNESCO Headquarters and the UNESCO International Institute for Educational Planning (IIEP); Union de Universidades de América Latina (UDUAL); Universities South Africa; and Associació Catalana d'Universitats Públiques (ACUP).

19.3 Digital Libraries

These are information repositories that include documents produced and published by the institutions that sponsor them.

19.3.1 OECD I-Library (http://www.oecd-ilibrary.org/education)

This is the online library of the Organisation for Economic Cooperation and Development (OECD) featuring its books, papers, and statistics and is the gateway to the OECD's analysis and data.

The education section includes an important number of books, reports, working papers, and articles on education and educational policy of the organization's member countries, with special emphasis on evaluation, analysis of results, and educational systems. They also emphasize their statistics and data series, including PISA (Programme for International Student Assessment) and TALIs (Teaching and learning International Survey—indicators).

19.3.2 IBEDOCS (http://www.ibe.unesco.org/)

This is a bibliographic database created by the Documentation Centre of the UNESCO International Bureau of Education. It contains bibliographic records on research and educational management throughout the world. The

Centre also has two other databases with national reports and curricular resources.

Resources collected at the Documentation Centre are part of the IBE knowledge base on curricula and education systems. This comprehensive set of specialized resources is at the service of capacity development and decision-making for quality education for everyone. The IBE documents and publications can be consulted through the **electronic catalogue** **UNESDOC** (https://unesdoc.unesco.org/explore/by-theme).

19.4 Education Plans and Policies Databases

A short list of resources containing information and documents on educational planning at the national level of a large number of countries around the world is provided hereunder.

19.4.1 Planipolis (http://planipolis.iiep.unesco.org/)

This is a resource created and maintained by the UNESCO's International Institute for Educational Planning. You can find national education plans and policies and key education frameworks and monitoring reports.

19.4.2 Eurydice (https://eacea.ec.europa.eu/national-policies/eurydice/home_en)

Eurydice is a network whose task is to explain how educational systems are organized in Europe and how they work. It publishes descriptions of national educational systems, studies on specific issues, indicators, and statistics in the field of education.

The reports show how countries address challenges at all educational levels: early childhood education and care, primary and secondary education, higher education, and adult learning.

19.5 Open Access Repositories

Unlike other disciplines, in education there are no thematic repositories that are used by the research community to massively deposit their scientific production in open access. Most researchers deposit their works (preprints or postprints) in multidisciplinary or institutional repositories. Here are two tools that make it possible to search simultaneously in most of these repositories:

19.5.1 BASE (https://www.base-search.net/)

Bielefeld Academic Search Engine (BASE) is a search motor that enables locating more than 140 million documents collected from more than 6,000 different sources. The majority of the resources (research reports, software, educational materials, etc.) are completely and freely available.

19.5.2 CORE (https://core.ac.uk/)

CORE is a tool created by the Joint Information Systems Committee (JISC) that allows collecting (and searching) more than 135 million documents of more than 4,600 providers of information: universities, research centers, scientific societies, etc. In most cases these are research reports (published or not).

19.6 Data Repositories

This section presents a nonexhaustive list of repositories containing data sets from studies, analyses, surveys, and educational management, at the micro- and macrolevels.

19.6.1 UIS.stat (http://data.uis.unesco.org/Index.aspx)

UNESCO Institute for Statistics Data Portal, which houses a powerful tool for displaying and downloading data and indicators produced or collected by themselves. The series can be customized according to three dimensions: indicator, country or region, and year.

19.6.2 EdStats (http://datatopics.worldbank.org/education/home)

This is a data portal that collects information about access, completion, learning, expenditures, policy, and equity in education, from data sets from other sources and their own, such as International learning assessments(PISA, TIMSS, PIRLS, PIAAC, and EGRA); regional learning assessments (SACMEQ, PASEC, and LLECE); World Bank databases, and household surveys such as LSMS, DHS, and MICS.

19.6.3 OECD Online Education Database (https://www.oecd.org/education/database.htm)

This is an access portal to different data sets produced by the OECD as part of its activities. Among the data sets are those corresponding to education at a Glance and Education GPS. Through the OECD I-Library you can also access more multiple sets of data (https://www.oecd-ilibrary.org/education/data/oecd-education-statistics_edu-data-en)

19.6.4 National Center for Education Statistics (https://nces.ed.gov/)

The National Center for Education Statistics (NCES) is a federal entity devoted to collecting and analyzing data related to education in the United States and other nations. NCES is located within the U.S. Department of Education and the Institute of Education Sciences. NCES has as its main mission to collect, collate, analyze, and report complete statistics on the condition of American education; conduct and publish reports; and review and report on education activities internationally.

A large number of data sets and analysis tools that can be used online or downloaded are available on its website.

19.7 Open Educational Resources (OER) Repositories

These are tools in which various types of materials are deposited, such as courses, curricular programs, didactic modules, student guides, evaluation tools, monographs, research articles, videos, streaming, podcasts, images, maps, multimedia and interactive materials, simulations, games, software, computer applications, mobile apps, or any other type of educational material and that are shared with open licenses so that they can be reused by the educational community free of charge.

There are a huge number of repositories that house such materials: institutional, research, thematic, national, and regional repositories.

We have again selected a couple of tools of collection, deposit, and creation of OER.

19.7.1 OER Commons (www.oercommons.org)

Created by the Institute for the Study of Knowledge Management in Education (ISKME) in 2007, OER Commons is a network with more than 50,000 educational resources available for anyone to use.

OER Commons has a search engine that enables locating educational materials using their metadata, which allows filtering the results according to parameters such as the educational level, the typology of the material, or the curricular area. Teachers and students can also label materials to enrich metadata and facilitate their discovery.

19.7.2 Merlot (www.merlot.org)

Merlot is a database of metadata that describes and permits locating all the learning materials that have been contributed to the repository by MERLOT members. As in other cases, it is a reusable material as it incorporates open licenses of the Creative Commons type.

References

Abascal, J., Barbosa, S.D., Nicolle, C., & Zaphiris, P. (2015). Rethinking universal accessibility: a broader approach considering the digital gap. *Universal Access in the Information Society*, 20, 1-4.

Adams Becker, S., Cummins, M., Davis, A., Freeman, A., Hall Geisenger, C., & Ananthanarayanan, V. (2017). *NMC Horizon Report: 2017 Higher Education Edition*. Austin, Texas: The New Medium Consortium.

Adams Becker, S., Freeman, A., Giesinger Hall, C., Cummins, M., & Yuhnke, B. (2016). *NMC/CoSN Horizon Report: 2016 K-12 Edition*. Austin, Texas: The New Medium Consortium.

Adarve, C., Castillo, D.A., Restrepo, E.J., & Villar-Vega, H. (2019). A review of virtual reality videogames for job-training applications. *Revista CINTEX*, 24(1), 64-70.

Aguaded, J.I., Vázquez-Cano, E. & Sevillano-García, M.L. (2013). MOOCs, ¿Turbocapitalismo de redes o altruismo educativo? In *SCOPEO Informe 2: MOOC: Estado de la situación actual, posibilidades, retos y futuro* (pp. 74-90). Salamanca: Universidad de Salamanca.

Ahir, K., Govani, K., Gajera, R., & Shah, M. (2020). Application on virtual reality for enhanced education learning, military training and sports. *Augmented Human Research*, 5(1), 7.

Alba, C. (2012). Aportaciones del Diseño Universal para el Aprendizaje y de los materiales digitales en el logro de una enseñanza accesible. En Actas del Congreso TenoNEEt. Retrieved from http://diversidad.murciaeduca.es/publicaciones/dea2012/docs/calba.pdf

Albion, P.R., Tondeur, J., Forkosh-Baruch, A., & Peeraer, J. (2015). Teachers' Professional Development for ICT Integration: Towards a Reciprocal Relationship between Research and Practice. *Education and Information Technologies*, 20(4), 655-673.

Alonso, T., & Gil-Cantero, F. (2019). El papel de la filosofía de la educación en la formación inicial docente. *Utopía y Praxis Latinoamericana*, 87, 27-42.

Alves, P., Miranda, L., & Morais, C. (2019). The Importance of Virtual Learning Environments in Higher Education. In *Computer-Assisted Language Learning: Concepts, Methodologies, Tools, and Applications* (pp. 109-131). New York: IGI Global.

Amador, G., Chávez, A.M., Alcaraz, N., Moy, N.A., Guzmán, J., & Tene, C. E. (2007). The role of tutors in self-directing the learning of the students of nursing. *Investigación y Educación en Enfermería*, 25(2), 52-59.

Amar, M. (2013). Educación y TIC en la sociedad del conocimiento. En J. Sánchez & J. Ruiz (Eds.), Recursos didácticos y tecnológicos en educación (pp. 15-23). Madrid: Editorial Síntesis.

American Library Association. (1989). *Presidential Committee on Information Literacy: Final Report..*

Anderson, T. (2016). Theories for Learning with Emerging Technologies. In G. Veletsianos (Ed.), *Emergence and Innovation in Digital Learning: Foundations and Applications* (pp. 35–50). Edmonton, AB: Athabasca University Press.

Area, M. & Adell, J. (2009). E-Learning: Enseñar y Aprender en Espacios Virtuales. In J. De Pablos. *Tecnología Educativa. La Formación del Profesorado en la Era de Internet* (pp. 391-424). Málaga: Ediciones Aljibe.

Armstrong, T. (2017). Inteligencias múltiples en el aula. Paidós Ibérica. Madrid.

Arnaiz, P. (2012). Escuelas eficaces e inclusivas: cómo favorecer su desarrollo. *Educatio Siglo XXI*, Vol. 30.1, pp. 25-44.

Arp, L. (1990). Information Literacy or Bibliographic Instruction: Semantics or Philosophy? *RQ* 30:46.

Association of College and Research Libraries. (2000). *Information Literacy Competency Standards for Higher Education.* Chicago, IL.: American Library Association.

Ayén, F. (2017). '¿Qué es la gamificación?'. *Iber: Didáctica de las Ciencias Sociales, Geografía e Historia,* 86, 7-15,

Azorín, C. & Arnaiz, P. (2013). Tecnología digital para la atención a la diversidad y mejora educativa. *Revista científica electrónica de Educación y Comunicación en la Sociedad del Conocimiento, 13*(1), 14-29.

Azuma, R. (1997). A Survey of Augmented Reality. *Presence: Teleoperators and Virtual Environments*, 4(6), 355-385.

Bacca, J., Baldiris, S. Fabregat, R., Graf, S & Kinshuk, J. (2014). Augmented Reality Trends in Education: A Systematic Review of Research and Applications, Educational *Technology & Society, 17* (4), 133-149.

Bakiri, N. (2016). Technology and Teacher Education: A Brief Glimpse of the Research and Practice that Have Shaped the Field. *TechTrends*, 60, 21-29.

Baldomero, M. (2015). The MECD Quality Certification Proposal of MOOC Courses. *International Journal of Educational Excellence*, 1 (2), 111-123.

Baldomero, M., & Leyva, J. J. (2016). EvalMOOC: A Pentadimensional Instrument of Improvement for the Quality Evaluation of MOOC. *International Journal of Educational Excellence*, 2(2), 15-33.

Baldwin, S.J., & Ching, Y.H. (2020). Guidelines for Designing Online Courses for Mobile Devices. *TechTrends*, 64(3), 413-422.

Ballock, E. (2019). Practitioner Research in a Changing Educator Preparation Landscape: Exploring Tensions and Reimagining Possibilities. *Journal of Practitioner Research, 4*(1), 1-5.

Barata, G. Gama, S., Jorge, J. & Gonçalves, D. (2017). 'Studying student differentiation in gamified education: A long-term study', *Computers in Human Behavior*, 71, pp 550-585

Barroso, J., & Gallego-Pérez, O. M. (2017). Producción de recursos de aprendizaje apoyados en Realidad Aumentada por parte de estudiantes de magisterio. *Revista de Educación Mediática y TIC (Edmetic), 6*(1), 23-38.

Barthes, R. (1957). *Mythologies*. París: la Seuil.

Batller, R. (2011). ¿De qué hablamos cuando hablamos de aprendizaje – servicio? *Crítica*. 61, 49–54.

Bauman, Z. (2015). *Modernidad Líquida*. México: Fondo de Cultura Económica.

Bawden, D. (2008). Origins and concepts of digital literacy. In C. Lankshear, M. Knobel (Eds.): *Digital literacies: Concepts, policies and practices*. New York: Peter Lang Publishing.

Bawden, D. (2001). Information and digital literacies: a review of concepts. Journal of Documentation 57, 218–259.

Bruce, C.S. 1997. *The seven faces of information literacy*. Auslib Press.

Bejinaru, R. (2019). Impact of Digitalization on Education in the Knowledge Economy. *Management Dynamics in the Knowledge Economy*, 7(3), 367-380.

Bell, F. (2011). Connectivism: Its Place in Theory-Informed Research and Innovation in Technology-Enabled Learning, *International Review of Research in Open and Distance Learning*, 12(3), 98-118.

Bello, A. & Merino, J. (2017). 'Socrative: A tool to dynamize the classroom', WPOM-Working Papers on Operations Management, 8, 72-75

Berelson, B. (1949). *Content Analysis in Communication Research*. Glencoe: The Free Press.

Berliner, D. (2002). Educational Research: The Hardest Science of All. *Educational Researcher.* 31(8), 18-20

Bicen, H. & Kocakoyun, S. (2017). 'Determination of university students' most preferred mobile application for gamification', *World Journal on Educational Technology*, 9, 18-23

Bielaczyc, K., & Collins, A. (1999). "Learning communities in classrooms: A reconceptualization of educational practice," in *Instructional-design theories and models: A new paradigm of instructional theory*, volume II, ed. C.M. Reiggeluth (New York, NY: Routledge), 269-292.

Biesta, G. (2015). On the Two Cultures of Educational Research, and How We Might Move Ahead: Reconsidering the Ontology, Axiology and Praxeology of Education. *European Educational Research Journal*, 14 (1), 11-22.

Billinghurst, M., Kato, H., & Poupyrev, I. (2001). The magicbook-moving seamlessly between reality and virtuality. *IEEE Computer Graphics and Applications, 21*, 6–8.

Bingimlas, K.A. (2009). Barriers to the successful integration of ICT in teaching and learning environments: A review of the literature. *Eurasia Journal of Mathematics, Science & Technology Education*, 5 (3), 235-245.

Blas, D., Morales, M.B., & López-Belmonte, J. (2019). Los retos didácticos de los MOOC en la sociedad de la información y el conocimiento. *Hekademos*, 27, 7-14.

Blaschke, L.M. (2012). Heutagogy and lifelong learning: A review of heutagogical practice and self-determined learning. *The International Review of Research in Open and Distributed Learning*, *13*(1), 56–71.

Blaschke, L.M., & Hase, S. (2015). Heutagogy, Technology, and Lifelong Learning for Professional and Part-Time Learners. In A. Dailey-Hebert & K. S. Dennis (Eds.), *Transformative Perspectives and Processes in Higher Education* (pp. 75–94). Cham: Springer International Publishing.

Bless, C., & Higson-Smith, C. (2004). *Fundamentals of Social Research Methods*. Lusaka: Juta.

Bogdanovic, Z., Barac, D., Jovanic, B., Popovic, S., & Radenkovic, B. (2014). Evaluation of Mobile Assessment in a Learning Management System. *British Journal of Educational Technology,* 45 (2), 231-244.

Bonk, C.J. (2009). *The world is open: how Web technology is revolutionizing education*. San Francisco, CA: Jossey-Bass.

Bonk, C.J., Wisher, R.A., & Nigrelli, M.L. (2004). Learning communities, communities of practice: Principles, technologies, and examples. In *Learning to collaborate, collaborating to learn*, eds. K. Littleton, D. Mieli, and D. Faulkner (Hauppauge, NY: Nova Science), 199-219.

Bressler, D.M. & Bodzin, A.M. (2013). A mixed methods assessment of students'flow experiences during a mobile augmented reality science game. *Journal of Computer Assisted Learning, 29*(6), 505-517.

Brown, A.L., & Campione, J.C. (1994). "Guided discovery in a community of learners" in *Classroom lessons: Integrating cognitive theory and classroom* practice, ed. K. McGilly (Cambridge, MA: The MIT Press), 229-270.

Bruce, C., Edwards, S., & Lupton, M. (2006). Six Frames for Information literacy Education: a conceptual framework for interpreting the relationships between theory and practice. *Innovation in Teaching and Learning in Information and Computer Sciences* 5, 1–18

Bruggeman, J. (2008). *Social Networks: An Introduction.* New York and London: Routledge.

Buabeng-Andoh, C. (2012). Factors Influencing Teachers' Adoption and Integration of Information and Communication Technology into Teaching: A Review of the Literature. *International Journal of Education and Development using Information and Communication Technology*, 8(1), 136-155.

Buckingham, D. (2003). *Media Education. Literacy, Learning and Contemporary Culture*. Oxford: Polity Press.

Buckley, P. & Doyle, E. (2017). 'Individualising gamification: An investigation of the impact of learning styles and personality traits on the efficacy of gamification using a prediction market', *Computers & Education*, 106, 43-55

Bunge, M. (1985). *Seudociencia e Ideología*. Madrid: Alianza.

Bunge, M. (1999). *Las Ciencias Sociales en Discusión. Una Perspectiva Filosófica*. Buenos Aires: EditorialSudamericana.

Bunge, M. (2010). *Matter and Mind: A Philosophical Inquiry*. Dordrecht: Springer Verlag.

Burden, R. & Williams, M. (1997). *Psychology for Language Teachers. A Social Constructivist Approach*. Cambridge: Cambridge University Press.

Burkle, M., & Magee, M. (2017). Virtual learning: videogames and virtual reality in education. In *Digital Tools for Seamless Learning* (pp. 325-344). New York: IGI Global.

Cabero J. (2006). Comunidades virtuales para el aprendizaje. Su utilización en la enseñanza. *Edutec. Revista Electrónica de Tecnología Educativa*, 20, 1-20.

Cabero J., Leiva J. J., Moreno N. M., Barroso J. & López-Meneses, E. (2016). *Realidad Aumentada y Educación. Innovación en contextos formativos*. Barcelona: Octaedro.

Cabero, J. & Barroso, J. (2015). Realidad Aumentada: posibilidades educativas. En J. Ruiz-Palmero, J. Sánchez-Rodríguez & E. Sánchez-Rivas (Edit.), *Innovaciones con tecnologías emergentes*. Málaga: Universidad de Málaga.

Cabero, J. & García, F. (Coords.) (2016). *Realidad aumentada. Tecnología para la formación*. Madrid: Síntesis.

Cabero, J., Barroso, J., Llorente, C. & Fernández-Martínez, M[a]. M. (2019). Educational Uses of Augmented Reality (AR): Experiences in Educational Science. *Sustainability, 11*, 4990; doi:10.3390/su11184990

Cabero, J., Fernández-Batanero, J.M. & Córdoba, M. (2016). Conocimiento de las TIC Aplicadas a las Personas con Discapacidades. Construcción de un Instrumento de Diagnóstico. *Revista Internacional de Investigación en Educación*, 88(7), 157-176.

Cabero, J., Vázquez-Cano, J. & López-Meneses, E. (2018). Uso de la Realidad Aumentada como Recurso Didáctico en la Enseñanza Universitaria. *Formación universitaria, 11*(1), 25-34. http://dx.doi.org/10.4067/S0718-50062018000100025

Carr, W. (1996). *Una teoría para la educación: hacia una investigación educativa crítica*. Madrid: Ediciones Morata.

Cassidy, S. (2016). Virtual Learning Environments as Mediating Factors in Student Satisfaction with Teaching and Learning in Higher Education. Journal *of Curriculum and Teaching*, 5(1), 113-123.

Cawood, S. & Fiala, M. (2008). *Augmented Reality: A Practical Guide*. Denver: Pragmatic Bookshelf.

Cebrián, G. & Junyent, M. (2015). Competencies in Education for sustainable development: Exploring the student teachers' view. *Sustainability*, *7*(3), 2768-2786.

Cebrián, M. (2011). Los centros educativos en la sociedad de la información y el conocimiento. En M. Cebrián & M. Gallego (Eds.), *Procesos educativos con TIC en la sociedad del conocimiento* (pp. 23-31). Madrid: Pirámide.

Çeker, E. & Özdamlı, F. (2017). 'What "Gamification" is and what it's not', European Journal of Contemporary Education, 6, 221-228.

Cerda, C., & Osses, S. (2012). Self-directed learning and self-regulated learning: two different concepts. *Revista Médica de Chile.* 140(11), 1504-1505.

Cerda, C., & Saiz, J.L. (2015). Self-directed learning in Chilean student teachers: A psychometric analysis. *Suma Psicológica.*, 22(2), 129-136.

Chang, H., Wu, K.& Hsu, Y. (2013). Integrating a mobile augmented reality activity to contextualize student learning of a socioscientific issue. *British Journal of Educational Technology, 44*, 3, 95-99.

Chang, Y.H., & Jen-ch'iang, L.I.U. (2013). Applying an AR Technique to Enhance Situated Heritage Learning in a Ubiquitous Learning Environment. *TOJET: The Turkish Online Journal of Educational Technology, 12*(3), 21-32.

Chang, Y., Long, J., & Hui, Y. (2013). Trends of Educational Technology Research: More Than a Decade of International Research in Six SSCI-Indexed Refereed Journals. *Education Technology Research Develeopment.* 61, 685–705.

Chen, C.-M., & Tsai, Y.-N. (2012). Interactive augmented reality system for enhancing library instruction in elementary schools. *Computers & Education, 59*(2), 638-652.

Chen, J.Q., & Gardner, H. (2012). Assessment of intelectual profile, a perspective from multiple-intelligences theory. En D. P. Flanagan, & P. L. Harrison (Eds), *Contemporary intellectual assessment. Theories, tests, and issues* (3rd ed.) (pp. 145-155). New York: Guilford Press.

Cheong, C. Cheong, F. & Filippou J. (2013). 'Quick Quiz: A Gamified Approach for Enhancing Learning', *Pacific Asia Conference on Information Systems*, p. 206

Cheung, A.C., & Slavin, R. (2016). How Methodological Features Affect Effect Sizes in Education. *Educational Researcher*, 45(5), 283–292.

Chiang, T.H., Yang, S.J., & Hwang, G.-J. (2014). Students' online interactive patterns in augmented reality-based inquiry activities. *Computers & Education, 78*, 97-108.

Christ, W.G., & Potter, W.J. (1998). Media Literacy, Media Education and the Academy. *Journal of Communication*, 48 (1), 5-15.

Chu, H.C., Hwang, G.J., & Tsai, C. C. (2010). A knowledge engineering approach to developing mindtools for context-aware ubiquitous learning. *Computers & Education, 54*(1), 289-297.

Clará, M. & Barberá, E. (2013). Learning Online: Massive Open Online Courses (MOOCs), Connectivism, and Cultural Psychology, *Distance Education*, 34(1), 129-136.

Clark, R.E. (1985). Confounding in Educational Computing Research. *Journal of Educational Computing Research.* 1 (2), 137-148.

Clark, R.E. & Salomon, C. (1977). Reexamining the Methodology of Research on Media and Technology in Education. *Review of Educational Research*, 47(1), 99-120.

Cloonan, A. (2019). Collaborative teacher research: integrating professional learning and university study. *The Australian Educational Researcher*, *46*(3), 385-403.

Cobb, L., and Steele, C. (2014). *Exploiting the fringe: flipping, microcredentials, and MOOCs*. Carrboro, NC: Tagoras.

Cobo, C. & Moravec, J. W. (2011). *Aprendizaje invisible. Hacia una nueva ecología de la educación*. Barcelona: Universitat de Barcelona.

Cohen, L., Manion, L., & Morrison, K. (2000). *Research Methods in Education*. Londres: Routledge.

Coll, C. (2001). *Las comunidades de aprendizaje y el futuro de la educación: el punto de vista del fórum universal de las culturas*. Universidad de Barcelona. Presented at the Simposio Internacional sobre Comunidades de Aprendizaje 2001.

Coll, C., Mauri, M. T., and Onrubia, J. (2008). Analyzing Actual Uses of ict in Formal Educational Contexts: A Socio-cultural Approach. *Revista Electrónica de Investigación Educativa*. 10(1).

Comer, J.P. (1980). *School Power*. New York: The Free Press.

Comisión de las Comunidades Europeas (2000) *Memorándum sobre el aprendizaje permanente*. 2000. SEC 1832, Brussels: Author.

Comisión de las Comunidades Europeas (2001) *Hacer realidad un espacio europeo del aprendizaje permanente.* 2001. COM 678 final, Brussels: Author.

Comisión de las Comunidades Europeas (2018) *Construyendo una Europa más fuerte: el papel de las políticas de juventud, educación y cultura*. 2018. COM 268 final, Brussels: Author.

Condliffe, E., & Shulman, L.S. (Ed) (1999). *Issues in Education Research: Problems and Possibilities*. San Francisco: National Academy of Education and Jossey-Bass Publishers.

Consejo de la Unión Europea (2009). Conclusiones del Consejo de 12 de mayo de 2009 sobre un marco estratégico para la cooperación europea en el ámbito de la educación y la formación («ET 2020»). C 119/02, Brussels: Author.

Consejo Europeo (2000). *Consejo Europeo de Santa María da Feira 19 y 20 de junio de 2000*. Conclusiones de la Presidencia. Santa María da Feira: Author.

Consejo Europeo (2002). *Consejo Europeo de Barcelona 15 y 16 de marzo de 2002*. Conclusiones de la Presidencia. 2002. Barcelona: Author.

Cooley, A. (2013). Qualitative research in education: The origins, debates, and politics of creating knowledge. *Educational Studies*, 49(3), 247-262.

Correa, J.M., Fernández, L., Gutiérrez, A., Losada, D., & Ochoa, B. (2015). Formación del Profesorado, Tecnología Educativa e Identidad Docente Digital. *Revista Latinoamericana de Tecnología Educativa*, 14(1), 45-56.

Correa, M. (2015). Flipping the foreign language classroom and critical pedagogies: A (new) old Trend. *Higher Education for the Future*. 2(2): 114-125.

Council of the European Union. (2018). Council Recommendation of 22 May 2018 on key competences for lifelong learning. *Official Journal of the European Union*.

Couros, A., & Hildebrant, K. (2016). Designing for Open and Social Learning. In G. Veletsianos (Ed.), *Emergence and Innovation in Digital Learning: Foundations and Applications* (pp. 143–161). Edmonton, AB: Athabasca University Press.

Crandall, P. G., Engler, R. K., Beck, D. E., Killian, S. A., O'Bryan, C. A., Jarvis, N., & Clausen, E. (2015). Development of an Augmented Reality Game to Teach Abstract Concepts in Food Chemistry. *Journal of Food Science Education*, 14(1), 18-23.

Cranmer, S., & Lewin, C. (2017). iTEC: conceptualising, realising and recognising pedagogical and technological innovation in European classrooms. *Technology, Pedagogy and Education*, *26*(4), 1–15.

Creswell, J.W. (2003). *Research Design: Qualitative, Quantitative, and Mixed Methods Approaches*. Thousand Oaks, CA: Sage.

Creswell, J W. (2012). *Educational Research: Planning, Conducting, and Evaluating Quantitative and Qualitative Research*. Boston: Pearson.

Cross, K.P. (1998). Why learning communities? Why now? *About campus*, 3: 4-11. [Stassen MLA. Student Outcomes: The Impact of Varying Living-Learning Community Models. Res. High. Educ. 2003; 44, 581-613.

Cuenca, J.M. & Martín, M. (2010). Virtual games in social science education. *Computers & Education*, 55 (3), 1336-1345,

Cumming, B. (2012). Revisiting Philosophical and Theoretical Debates in Contemporary Educational Research and Major Epistemological and Ontological Underpinnings. *ERIC Number*: ED537463. Online Submission.

Dabbagh, N., & Kitsantas, A. (2012). Personal Learning Environments, social media, and self-regulated learning: A natural formula for connecting formal and informal learning. *Social Media in Higher Education, 15*(1), 3–8.

De Salces, F.J.S., Martín, M.T., Perea, L.G., & Madrid, I. (2014). Involucración de personas con discapacidad en proyectos tecnológicos de I+ D+ i: el caso de APSIS4all. *Revista Española de Discapacidad (REDIS), 2*(2), 121-144.

Delgado, A.I. (2013). Massive Open Online Course (MOOC), ¿Un Sustituto Irreversible de Moodle? *Revista Electrónica Conocimiento Libre y Licenciamiento*, 6(4), 55-61.

Delgado-Algarra, E.J., Montes-Navarro, M. A. & Lorca-Marín. A. A. (2018) 'La integración de las didácticas específicas a través de los videojuegos: revisión de contenidos escolares en SimCity' Cive2018, *VI Congreso Internacional de Videojuegos y Educación*, 12-14 de septiembre, 2018.

Delgado-Algarra, E.J. (2014). 'Memoria histórica y concienciación cívica a través del videojuego This War of Mine'. *Clio: History and History Teaching*, 40,

Delgado-Algarra, E.J. (2019). 'Enseñanza de las historia y compromise ciudadano a través de los videojuegos Civilization VI y Stardew Valley: cómo seleccionar e integrar los videojuegos en el aula'. *Clio*, 44.

Delgado-Vázquez, A. Vázquez-Cano, E., Belando-Montoro, M. R. & López-Meneses, E. (2019). Análisis bibliométrico del impacto de la investigación educativa en diversidad funcional y competencia digital: Web of Science y Scopus. *Aula Abierta*, 48 (2), 147-156.

Delors, J., International Commission on Education for the Twenty-first Century. 1996. Learning: the treasure within; report to UNESCO of the International Commission on Education for the Twenty-first Century (highlights).

De-Marcos, L., Garcia-Lopez, E. & Garcia-Cabot, A. (2016). 'On the effectiveness of game-like and social approaches in learning: Comparing educational gaming, gamification & social networking', *Computers & Education*, 95, 99-113

De-Marcos, L. García-López, & E. García-Cabot, A. (2016). 'On the effectiveness of game-like and social approaches in learning: Comparing educational gaming, gamification & social networking', *Computers & Education*, 95, 99-113

Denzin, N. & Lincoln, Y. (Eds.). (2008). *Collecting and Interpreting Qualitative Materials*. Thousand Oaks, CA: Sage.

Díaz-Posada, L.E., Varela-Londoño, S.P, & Rodríguez-Burgos, L.P. (2017). Inteligencias múltiples e implementación del currículo: avances, tendencias y oportunidades. *Revista de Psicodidáctica*, 22(1), 69–83.

Dicheva, D., Dichev, C., Agre, G., & Angelova, G. (2015). 'Gamification in education: A systematic mapping study', *Journal of Educational Technology & Society*, 18, pp. 75-88

Domínguez, A., Saenz-de-Navarrete, J., De-Marcos, L., Fernández-Sanz, L., Pagés, C., & Martínez-Herráiz, J.J. (2013). 'Gamifying learning experiences: Practical implications and outcomes'. *Computers & Education*, 63, pp. 380-392

Driscoll, M.P., and Vergara, A. (1997). New technologies and their impact on the education of the future [Nuevas Tecnologías y su impacto en la educación del futuro]. *Pensamiento Educativo*. 21(2): 81-99.

DuFour, R. (2004). What Is a "Professional Learning Community"? *Educ. Leadership* 61: 6-11.

Dunleavy, M., Dede, C., & Mitchell, R. (2009). Affordances and limitations of immersive participatory augmented reality simulations for teaching and learning. *Journal of Science Education and Technology, 18* (1), 7-22. doi:10.1007/s10956- 008-9119-1.

Dziuban, C., Graham, C. R., Moskal, P. D., Norberg, A., & Sicilia, N. (2018). Blended learning: the new normal and emerging technologies. *International Journal of Educational Technology in Higher Education*, 15(1), 3.

Eco, U. (1979). *The Role of the Reader. Explorations in the Semiotics of Texts*, Bloomington: Indiana University Press.

Eisenberg, M.B., & Berkowitz, R.E. (1990). *Information problem-solving: the Big Six Skills approach to library & information skills instruction*. Norwood, NJ.: Ablex.

Elboj, C., Valls, R., & Fort, M. (2000). Comunidades de aprendizaje. Una práctica educativa para la sociedad de la información. *Cult. Educ.*, 12, 129-141.

Ellis, A. (2005). What Research Is of Most Worth? In VV.AA. *Research on Educational Innovations* (4th Ed.), Inc. Larchmont, NY.

Elmqaddem, N. (2019). Augmented reality and virtual reality in education. Myth or reality?. *International Journal of Emerging Technologies in Learning*, 14(03), 234-242.

Ely, D.P. (1999). Conditions that Facilitate the Implementation of Educational Technology Innovations. *Educational Technology*, 39 (6), 23-27.

Engeström, Y. (1987). *Learning by expanding: An activity-theoretical approach to developmental research.* Helsinki: Orienta-Konsultit.

Engeström, Y. (2007). Enriching the Theory of Expansive Learning: Lessons From Journeys Toward Coconfiguration. *Mind, Culture, and Activity, 14*(1–2), 23–39.

Escobar, J.P., Arroyo, R., Benavente, C., Díaz, R., Garolera, M., Sepúlveda, A., Urzúa, D., & Veliz, S. (2016). Requisitos, retos y oportunidades en el context del desarrollo de nuevas tecnologías con niños para niños con discapacidad. *Revista Nacional e Internacional de Educación Inclusiva*, 9(3), 213–219.

Etchegaray, M.C., Guzmán, M.D., & Duarte, A.M. (2017). Diseño de un recurso multimedia on line basado en Inteligencias Múltiples. *Campus Virtuales*, 6(1), 51-65.

European Commission. (2010). *A Digital Agenda for Europe: Communication from the Commission to the European Parliament, the Council, the European Economic and Social Committee and the Committee of the Regions*. Brussels: European Commission.

European Parliament and the Council. (2006). Recommendation of the European Parliament and the Council of 18 December 2006 on key competences for lifelong learning, 2006/962/EC. *Official Journal of the European Union* L394/310.

EUROSTAT (2019). *Adult learning statistics*. Retrieved from https://ec.europa.eu/eurostat/statistics-explained/index.php/Adult_learning_statistics

Fabris, C.P., Rathner, J.A., Fong, A.Y., & Sevigny, C.P. (2019). Virtual Reality in Higher Education. *International Journal of Innovation in Science and Mathematics Education*, 27(8), 1-9.

Fasce, E., Pérez, C., Ortiz, L., Parra, P., Ibáñez, P., & Matus, O. (2013). Relationship between self-directed learning and value profile in Chilean medical students. *Revista Médica de Chile*. 141(1): 15-22.

Feibleman, J.K. (1983). Pure Science, Applied Science, and Technology: An Attempt at Definitions. En C. Mitcham & R. Mackey (Eds.), *Philosophy and Technology* (pp. 33-41). New York: The Free Press.

Feinberg, W. (2012). Critical Pragmatist and the Reconnection of Science and Values in Educational Research. *European Journal of Pragmatism and American Philosophy*, 4 (1), 222-240.

Ferguson, R. (1999). The Mass Media and the Education of Students in a Democracy: Some Issues to Consider. *The Social Studies*, 90 (6), 257-261.

Fernández Batanero, J.M., Román, P., & El Homrrani, M. (2017). TIC y discapacidad. Conocimiento del profesorado de educación primaria en Andalucía. *Aula Abierta*, 46, 65-72.

Fernández Batanero, J.Mª, Reyes Rebollo, M.Mª., & El Homran, M. (2018). TIC y discapacidad. Principalesbarreras para la formación del profesorado. EDMETIC, Revista de EducaciónMediática y TIC, *7*(1), 1-25.

Fernández Batanero, J.M., Cabero, J., & López Meneses, E. (2018). Knowledge and degree of training of primary education teachers in relation to ICT taught to students with disabilities. *British Journal ofEducational Technology.* doi:10.1111/bjet.12675

Fernández-Batanero, J.M. (2018). Tic y la discapacidad. Conocimiento del profesorado de Educación Especial. *Revista Educativa Hekademos, 24*, 19-29.

Fernández-Batanero, J.M., & Rodríguez-Martín, A. (2017). TIC y diversidad funcional: conocimiento del profesorado. *European Journal of Investigation in Healh, Psychology and Education, 7*(3), 157-175.

Fernández-Ramírez, B. (2014). En Defensa del Relativismo: Notas Críticas desde una Posición Construccionista. *Aposta, Revista de Ciencias Sociales.* 60,1-36.

Fernández-Robles, B. (2018). La utilización de objetos de aprendizaje de realidad aumentada en la enseñanza universitaria de educación primaria. *International Journal of Educational Research and Innovation (IJERI)*, 9, 90-104.

Ferrandis, M.V., Grau, C., & Fortes, M.C. (2010). El profesorado y la atención a la diversidad en la ESO. *Revista de EducaciónInclusiva*, 3(2), 11-28.

Ferrari, A. (2012). *Digital Competence in Practice: An Analysis of Frameworks*. doi: 10.2791/82116.

Ferrari, A. (2013). *DIGCOMP: A Framework for Developing and Understanding Digital Competence in Europe*. [Seville]: Institute for Prospective Technological Studies. doi: 10.2788/52966.

Ferreiro, R., & De Napoli, A. (2006). Un Concepto Clave para Aplicar Exitosamente las Tecnologías de la Educación: Los Nuevos ambientes de Aprendizaje. *Revista Panamericana de Pedagogía*, 8, 121-154.

Ferrer-Torregrosa, J., Torralba, J., Jimenez, M., García, S., & Barcia, J. (2015). ARBOOK: Development and assessment of a tool based on augmented reality for anatomy. *Journal of Science Education and Technology, 24*(1), 119e124.

Fidalgo, A., Sein, M.L., & García-Peñalvo, F.J. (2016). From massive access to cooperation: lessons learned and proven results of a hybrid xMOOC/cMOOC pedagogical approach to MOOCs. International Journal of Educational Technology in Higher Education, 13(1), 24.

Findlay-Thompson, S., & Mombourquette, P. (2014). Evaluation of a flipped classroom in an undergraduate business course. Business Education and Accreditation. 6(1): 63-71.

Fitzgerald, G., Koury, K., & Mitchem, K. (2008). Research on computer-mediated instruction for students with high incidence disabilities. *Journal of Educational Computing Research*, *38*(2), 201–233.

Flecha, J.R., & Puigvert L. (2002). Las comunidades de aprendizaje: una apuesta por la igualdad educativa. REXE: *Revista de estudios y experiencias en educación* 1, 11–20.

Flecha, R., & Puigvert, L. (2005). Formación del profesorado en las comunidades de aprendizaje. *Revista Colombiana de Educación*. 48, 12–36.

Flick, U. (2016). Challenges for a New Critical Qualitative Inquiry: Introduction to the Special Issue. *Qualitative Inquiry*. 5, 1–5.

Flyvbjerg, B. (2011). Case Study. En N. K. Denzin, & Y. S. Lincoln (Ed). *The Sage Handbook of Qualitative Research*. Londres: Sage.

Fombona, J., Pascual, M.J., & Madeira, M.F. (2012). Realidad aumentada, una evolución de las aplicaciones de los dispositivos móviles. *Píxel-Bit. Revista de Medios y Educación, 41*, 197-210.

Forsythe, E. (2015). Improving Assessment in Japanese University EFL Classes: A Model for Implementing Research-Based Language Assessment Practices. *21st Century Education Forum*, 10, 65–73.

Galán, A., Ruiz-Corbella, M., & Sánchez Mellado, J.C. (2014). Repensar la Investigación Educativa: De las Relaciones lineales al Paradigma de la Complejidad. *Revista Española de Pedagogía*, 258, 281-298.

Gallego, D.J. & Alonso, C. (2000). *La Informática en la Práctica Docente*. Madrid: UNED/Editorial Edelvives.

Galway, L.P., Corbett, K.K., Takaro, T.K., Tairyan, K., & Frank, E. (2014). A Novel Integration of Online and Flipped Classroom Instructional Models in Public Health Higher Education. BMC Medical Education. 14, 181, 1–9.

Gannon-Leary, P., & Fontainha, E. (2007). Communities of Practice and Virtual Learning Communities: Benefits, Barriers and Success Factors. *eLearning Papers*, 5, 1–14.

García, B., Tenorio, G., & Ramírez, M. (2015). Retos de automotivación para el involucramiento de estudiantes en el movimiento educativo abierto con MOOC. *RUSC-Universities and Knowledge Society Journal*, 12(1), 91-104.

García, T., Fernández, E., Vázquez, A., García, P., & Rodríguez, C. (2018). El Género y la Percepción de las Inteligencias Múltiples. *Revista Psicología Educativa* 24(1) 31-37

García-Barrera, A. (2013). The reverse classroom: changing the response to students' needs [El aula inversa: cambiando la respuesta a las necesidades de los estudiantes]. *Avances en Supervisión Educativa*. 19.

García-Rangel, M., & Quijada-Monroy, V.C. (2015). The inverted classroom and other strategies with the use of ICT. Learning experience with teachers [El Aula invertida y otras estrategias con uso de TIC. Experiencia de aprendizaje con docentes]. Memoria 2015 Universidad Nacional Autónoma de México (UNAM) Available at: http://somece2015.unam.mx/MEMORIA/57.pdf

Gardner, H. (2005). Inteligencias múltiples. *Revista de Psicología y Educación*, 1(1), 17-26.

Gay, R.L., Mills, G.E., & Airasian, P.W. (2011). *Educational Research: Competencies for Analysis and Application*. Upper Saddle River: Merrill.

Gilbert, J. (2019). Is Complexity Thinking a Useful Frame for Change-Oriented Educational Research?. In *Educational Research in the Age of Anthropocene* (pp. 259-284). New York: IGI Global.

Gil-Cantero, F., & Reyero, D. (2014). La Prioridad de la Filosofía de la Educación sobre las Disciplinas Empíricas de la Investigación Educativa. *Revista Española de Pedagogía,* 258, 263-280.

Gilster, P. (1997). *Digital literacy*. New York: John Wiley & Sons Inc.

Gisbert, M., & Lázaro, J.L. (2015). Professional development in teacher digital competence and improving school quality from the teachers' perspective: a case study. Journal of New Approaches in Educational Research, 4(2), 115. https://doi.org/10.7821/naer.2015.7.123

Gómez Contreras, J.L., & Camargo, D.A. (2019). Virtual Learning Environments In Accounting Education. *Economy & Business Journal*, 13(1), 224-231.

Gómez Galán, J. (1999). *Tecnologías de la Información y la Comunicación en el Aula*. 2 vols. Madrid: Seamer.

Gómez Galán, J. (2001). Aplicaciones Didácticas y Educativas de las Tecnologías RIV (Realidad Infovirtual) en Entornos Telemáticos. En VV.AA. *Actas del XIII Congreso Internacional de Ingeniería Gráfica: Eliminando Fronteras entre lo Real y lo Virtual*. (pp. 29-43). Badajoz: AEIA-UEX.

Gómez Galán, J. (2002). Education and Virtual Reality. En N. Mastorakis (ed.). *Advances in Systems Engineering, Signal Processing and Communications*. (pp. 345-350). Nueva York: WSEAS Press

Gómez Galán, J. (2003). *Educar en Nuevas Tecnologías y Medios de Comunicación*. Sevilla-Badajoz: Fondo Educación CRE.

Gómez Galán, J. (2007). Los Medios de Comunicación en la Convergencia Tecnológica: Perspectiva Educativa. *Comunicación y Pedagogía*, 221, 44-50.

Gómez Galán, J. (2011). New Perspectives on Integrating Social Networking and Internet Communications in the Curriculum. *eLearning Papers*, 26, 1-7.

Gómez Galán, J. (2012). Corrientes de Investigación en Tecnología Educativa. En J. Gómez Galán & G. Lacerda (Coords.). *Informática e Telemática na Educação*. Volumen I. *As Tecnologias de Informação e Comunicação na Educação*. (pp. 105-159). Brasilia: Liber Livro Editora/Universidade de Brasilia.

Gómez Galán, J. (2013). Realidad Virtual en la Arqueología y el Arte: Orientaciones Didácticas y Formativas. En VV.AA. *Arte y Sociedad: Bellas Artes y Sociedad Digital*. (pp. 5-17). Málaga: FUAIG

Gómez Galán, J. (2014a). Educación y Globalización: Desarrollos Pedagógicos Innovadores en el Contexto de la Cultura Postmoderna. En J. C. Martínez Coll (Ed.), *Educación, Cultura y Desarrollo*. Málaga: Fundación Universitaria Andaluza Inca Garcilaso.

Gómez Galán, J. (2014b). Transformación de la Educación y la Universidad en el Postmodernismo Digital: Nuevos Conceptos Formativos y Científicos. En F. Durán (Ed.). *La Era de las TIC en la Nueva Docencia* (pp. 171-182). Madrid: McGraw-Hill.

Gómez Galán, J. (2014c). Taxonomía y Metodología Científica de la Tecnología Educativa: Dimensiones Intraeducacional y Supraeducacional. In *Proceedings of International Congress SEECI 2014*. Madrid: UCM

Gómez Galán, J. (2015a). Introduction: A New Educational Research to Society, Citizens and Human Development. En J. Gómez Galán, E. López Meneses & Martín, A. H. (Eds.). (2015). *Advances and Innovations in Educational Research* (pp. 7-12). Cupey: UMET Press.

Gómez Galán, J. (2015b). Introduction: Social Sciencies Research in the ICT Age. En J. Gómez Galán, E. López Meneses & L. Molina (Eds.). *Research Foundations of the Social Sciences* (pp. 6-10). Cupey: UMET Press.

Gómez Galán, J. (2017a). Nuevos estilos de enseñanza en la era de la convergencia tecno-mediática: hacia una educación holística e integral. *IJERI: International Journal of Educational Research and Innovation*, 8, 60-78.

Gómez Galán, J. (2017b). Educational Architecture and Emerging Technologies: Innovative Classroom-Models. *Hekademos,* 22, 7-18.

Gómez Galán, J. (2017c). Educational Research and Teaching Strategies in the Digital Society: A Critical View. In E. López Meneses, F. Sirignano, M. Reyes, M. Cunzio & J. Gómez Galán. *European Innovations in Education: Research Models and Teaching Applications* (pp. 105-119) Seville: AFOE

Gómez Galán, J. (2017d). Interacciones Moodle-MOOC: presente y futuro de los modelos de e-learning y b-learning en los contextos universitarios. *EccoS Revista Científica*, 44, 241-257.

Gómez Galán, J. (2020). MOOC courses in the context of distance education: a new pedagogical approach. In R.V. Nata (ed.). *Progress in Education.* New York: Nova Science Publishers.

Gómez Galán, J. (Ed.). (2016). *Educational Research in Higher Education: Methods and Experiences.* Aalborg: River Publishers.

Gómez Galán, J., & Mateos, S. (2002). Retos Educativos en la Sociedad de la Información y la Comunicación. *Revista Latinoamericana de Tecnología Educativa*, 1 (1), 9-23.

Gómez Galán, J., & Mateos, S. (2004). Design of Educational Web Pages. *European Journal of Teacher Education.* 27 (1), 99-104.

Gómez Galán, J., & Sáenz del Castillo, A.A. (Eds.) (2000). *Nuevas Tecnologías Aplicadas a la Educación.* Badajoz: Universitas Editorial.

Gómez Galán, J., & Sirignano, F. (Eds.) (2016). *Theory and Practice in Educational Research.* Nápoles: Edizioni Università degli Studi Suor Orsola Benincasa.

Gómez Galán, J., Lázaro, C., Martínez López, J.A., & López Meneses, E. (2020). Measurement of the MOOC phenomenon by pre-service teachers: A descriptive case study. *Education Science*, 10(8), 207.

Gómez Galán, J., Martín, A.H., Bernal, C., & López Meneses, E. (2019). *MOOC Courses and the Future of Higher Education: A New Pedagogical Framework*. Aalborg: River Publishers.

Gómez-Galán, J. (2020). Media Education in the ICT Era: Theoretical Structure for Innovative Teaching Styles. *Information, 11*(5), 276.

Gonzalez, A., & Farnós, J. (2009). Usabilidad y accesibilidad para un e-learning inclusivo. Revista de Educación Inclusiva, 2 (1), 49-60.

González, R., & Gutiérrez, A. (2017). Competencias Mediática y Digital del profesorado e integración curricular de las tecnologías digitales. *Revista Fuentes, 19*(2), 57-67.

Gorard, S., & Taylor, C. (2004) *Combining Methods in Educational and Social Research*. Londres: Open University Press.

Gore, J.M. (2017). Reconciling educational research traditions. *The Australian Educational Researcher*, *44*(4-5), 357-372.

Goshevski, D., Veljanoska, J., & Hatziapostolou, T. (2017). A review of gamification platforms for higher education, *Proc. In 8th Balkan Conference in Informatics*, Skopje, Macedonia

Goska, R.E., & Ackerman, P.L. (1996). An Aptitude–Treatment Interaction Approach to Transfer within Training. *Journal of Educational Psychology*, 88 (2), 249.

Grabe, M., & Grabe, C. (2007). *Integrating technology for meaningful learning (5th ed.)*. Boston, MA: Houghton Mifflin.

Graham, P. (2007). Improving Teacher Effectiveness through Structured Collaboration: A Case Study of a Professional Learning Community. *RMLE Online* 31, 1–17.

Green, B. (2010). Knowledge, the Future, and Education(al) Research: A New-Millennial Challenge. *The Australian Educational Researcher*. 37(4), 43-62.

Greenwood, D.J. & Levin, M. (2006). *Introduction to Action Research: Social Research for Social Change*. Thousand Oaks, CA: Sage.

Guglielmino, L.M. (1978). *Development of the self-directed learning readiness scale*. PhD. thesis. Unifersity of Georgia, United States.

Gunasinghe, A., Hamid, J.A., Khatibi, A., & Azam, S.M. (2018). Does the lecturer's innovativeness drive VLE adoption in higher education institutes?. *Journal of Information Technology Management*, 10(3), 20-42.

Gutiérrez, K.D., Baquedano López, P., & Tejeda, C. (1999). Rethinking diversity: Hybridity and hybrid language practices in the third space. *Mind, Culture, and Activity*, *6*(4), 286–303.

Guzmán, V.F., & Vila, J.R. (2011). Open educational resources and use of internet in higher education: opencourseware project. Edutec-e, Revista Electrónica de Tecnología Educativa. 38, a182.

Haag, S., Cummings, M., & McCubbrey, D.J. (2004). *Management information systems for the information age (4th Ed.).* New York: McGraw-Hill.

Hakkarainen, K., & Paavola, S. (2009). Toward a trialogical approach to learning. In B. Schwarz, T. Dreyfus, & R. Hershkowitz (Eds.), *Transformation of knowledge through classroom interaction* (pp. 65–80). London; New York: Routledge.

Hall, S. (1977). Culture, the Media and the Ideological Effect. En J.Curran, M. Gurevitch & J. Woollacott (eds.) *Mass Communication and Society.* (pp. 315-348). London: Edward Arnold.

Hall, S., & Whannel, P. (1964). *The Popular Arts*. London: Hutchinson.

Hammersley, M. (2004). Action Research: A Contradiction in Terms? *Oxford Review of Education*, 30 (2), 165-181.

Hamutoglu, N.B., Gemikonakli, O., Duman, I., Kirksekiz, A., & Kiyici, M. (2020). Evaluating students experiences using a virtual learning environment: satisfaction and preferences. *Educational Technology Research and Development*, 68(1), 437-462.

Han, J., Jo, M., Hyun, E., & So, H. (2015). Examining young children's perception toward augmented reality-infused dramatic play. *Education Technology Research Development, 63*, 455-474.

Hargrave, C.P., Simonson, M.R., & Thompson, A.D. (1996). *Educational Technology: A Review of the Research.* Ames: Iowa State University.

Hart, A. (1991). *Understanding the Media.* London and New York: Routledge.

Haynes, N.M., Comer, J.P., & Hamilton-Lee, M. (1988). The School Development Program: A Model for School Improvement. *J. Negro Educ.* 57, 11–21.

Hegarty, J., Bostock, S., & Collins, D. (2000). Staff development in information technology for special needs: a new distance-learning course at Keele University. *British Journal of Educational Technology*, *31*(3), 199–212.

Heick, T. (2018). Exactly How To Teach With Video Games In The Classroom. *Teachthought*. Retrieved from https://www.teachthought.com

Hersh, M. (2017). Classification framework for ICT-based learning technologies for disabled people. *British Journal of Educational Technology, 48*(3), 768–788.

Hew, K.F., Lan, M., Tang, Y., Jia, C., & Lo, C.K. (2019). Where is the "theory" within the field of educational technology research?. *British Journal of Educational Technology*, *50*(3), 956-971.

Hill, J., Pettit, J., & Dawson, G. (1995). *Schools as learning communities, A Discussion Paper* NSW Department of School Education, Sydney.

Hjorth, L., & Hinton, S. (2019). *Understanding social media.* SAGE Publications Limited.

Hoadley, C., & Kali, Y. (2019). Five waves of conceptualizing knowledge and learning for our future in a networked society. In Y.Kali, A. Baram-Tsabary, A., Schejter (Eds.). *Learning in a networked society: Spontaneous and designed technology enhanced learning communities.* Singapore: Springer

Hobbs, R., & Jensen, A. (2009). The Past, Present, and Future of Media Literacy Education. *Journal of Media Literacy Education*, 1 (1), 1-11.

Holmberg, J. (2017). Applying a conceptual design framework to study teachers' use of educational technology. *Education and Information Technologies*, *22*(5), 2333-2349.

Hood, K. (2017). Telling Active Learning Pedagogies Apart: from theory to practice. *Journal of new Approaches in Educational Research, 6*(2), 144-152.

Hord, S.M. (2008). Evolution of the professional learning community. *The Learning Professional* 29, 10.

Huang, K.T., Ball, C., Francis, J., Ratan, R., Boumis, J., & Fordham, J. (2019). Augmented versus virtual reality in education: an exploratory study examining science knowledge retention when using augmented reality/virtual reality mobile applications. *Cyberpsychology, Behavior, and Social Networking*, 22(2), 105-110.

Hung, P.H., Hwang, G. J., Lin, Y. F., Wu, T.H., & Su, I.H. (2013). Seamless connection between learning and assessmentapplying progressive learning tasks in mobile ecology inquiry. *Educational Technology & Society, 16*(1), 194-205.

Hussain, A.J., Morgan, S., & Al-Jumeily, D. (2011). How Does ICT Affect Teachings and Learning within School Education. *In Developments in E-systems Engineering (DeSE),* pp. 250-254. New York: IEEE.

Huttar, C. M., & Brintzenhofeszoc, K. (2020). Virtual reality and computer simulation in social work education: A systematic review. *Journal of Social Work Education*, 56(1), 131-141.

Infante-Moro, A., Infante-Moro, J.C., Torres-Díaz, J.C., Martínez-López, f.J. (2017). Los MooC como sistema de aprendizaje en la Universidad de Huelva (UhU). IJERI: International Journal of Educational Research and Innovation, 7, 13–24.

Ingram, N.R. (2016). Time past: impacts of ICT on the pedagogic discourse in the Interactive project. *Technology, Pedagogy and Education, 25*(1), 1–18.

ISFE (2017). *The New Faces of Gaming. Report*. New York: Author.

Isin, E.F., & Ruppert, E.S. (2020). *Being digital citizens*. Lanham, Maryland: Rowman & Littlefield Publishers.

Istenic, A., & Bagon, S. (2014). ICT supported learning for inclusion of people with special needs: review of seven educational technology journals, 1970-2011. *British Journal of Educational Technology*, *45*(2), 202–230.

Jacobson, M.J., Levin, J.A., & Kapur, M. (2019). Education as a complex system: Conceptual and methodological implications. *Educational Researcher*, *48*(2), 112-119.

Janicki, T.C. & Peterson, P.L. (1981). Aptitude-Treatment Interaction Effects of Variations in Direct Instruction. *American Educational Research Journal*, 18 (1), 63-82.

Januszewski, A., & Molenda, M. (Eds.). (2013). *Educational Technology: A Definition with Commentary*. New York and London: Routledge.

Jenkins, H., Clinton, K., Purushotma, R., Robison, A.J., & Weigel, M. (2006). *Confronting the Challenges of Participatory Culture. Media Education for the 21st Century*. Londres-Cambridge, MA: The MIT Press.

Jiménez-Palacios, J. M. Cuenca. (2017). Libertus. *Iber: Didática de las Ciencias Sociales, Geografía e Historia*, 86, 41–44.

Johnson, L., Adams Becker, S., Cummins, M., Estrada, V., Freeman, A., & Hall, C. (2016). *NMC Horizon Report: 2016 Higher Education Edition*. Austin, Texas: The New Media Consortium.

Johnson, R.B., & Christensen, L.B. (2007). *Educational Research: Quantitative, Qualitative, and Mixed Approaches*. Thousand Oaks, CA: Sage.

Johnson, T.R., Lyons, R., Kopper, R., Johnsen, K.J., Lok, B.C. & Cendan, J.C. (2014). Virtual Patient Simulations and Optimal Social Learning Context: A Replication of an Aptitude–Treatment Interaction Effect. *Medical Teacher*, 36 (6), 486-494.

Jokonya, O. (2016). The Significance of Mixed Methods Research in Information Systems Research. *MWAIS 2016 Proceedings.* Paper 20. Retrieved from http://aisel.aisnet.org/mwais2016/20

Judson, E., & Sawada, A. (2002). 'Learning from past and present: Electronic response systems in college lecture halls', *Journal of Computers in Mathematics and Science Teaching*, 21, 167–181

Juhaňák, L., Zounek, J., Záleská, K., Bárta, O., & Vlčková, K. (2019). The relationship between the age at first computer use and students' perceived competence and autonomy in ICT usage: A mediation analysis. *Computers & Education, 141* https://doi.org/10.1016/j.compedu.2019.103614.

Kahoot Official Website: 'About us'. Available at https://kahoot.uservoice .com/knowledgebase/articles/464890-who-and-what-is-behind-kahoot [Accessed April 11, 2018].

Kahoot Official Website: 'About us'. Available at https://kahoot.com/compa ny/ [Accessed April 11, 2018].

Kali Y., Sagy O., Benichou, M., Atias, O., & Levin-Peled (2019). Teaching expertise reconsidered: The Technology, Pedagogy, Content and Spaces (TPeCS) knowledge framework. *British Journal of Educational Technology 50*(5), 2162–2177

Kali, Y., Baram-Tsabari, A., & Schejter A. (Eds.) (2019). *Learning in a networked society: Spontaneous and designed technology enhanced learning communities*. New York: Springer.

Kamińska, D., Sapiński, T., Wiak, S., Tikk, T., Haamer, R.E., Avots, et al. (2019). Virtual reality and its applications in education: Survey. *Information*, 10(10), 318.

Kanfer, R. & Ackerman, P. L. (1989). Motivation and Cognitive Abilities: An Integrative/Aptitude-Treatment Interaction approach to Skill Acquisition. *Journal of Applied Psychology*, 74 (4), 657.

Kang, N. (2015). The comparison between regular and flipped classrooms for EFL Korean adult learners. *Multimedia-Assisted Language Learning*. 18(3): 41-72.

Kang, S., & Kang, S. (2019). The study on the application of virtual reality in adapted physical education. *Cluster Computing*, 22(1), 2351–2355.

Kaplan, D. (2015). The Future of Quantitative Inquiry in Education: Challenges and Opportunities. En M. J. Feuer, A. I. Berman, & R. C. Atkinson (Eds). *Past aPrologue: The National Academy of Education at 50. Members Reflect.* (pp. 109-117). Washington, D.C.: National Academy of Education.

Kellogg, K. (1999). *Learning Communities*. ERIC Digest. Washington, DC: ERIC Clearinghouse on Higher Education. (ED430512).

Kenttälä, V., & Kankaanranta, M. (2020). Building ground for flexible use of educational technology. *Information Technology, Education and Society*, *17*(1), 21-40.

Kerawalla, L., Luckin, R., Seljeflot, S., & Woolard, A. (2006). "Making it real": Exploring the potential of augmented reality for teaching primary school science. *Virtual Reality, 10*(3-4), 163-174.

Kerimbayev, N., Nurym, N., Akramova, P., & Abdykarimova, S. (2020). Virtual educational environment: Interactive communication using LMS Moodle. *Education and Information Technologies*, 25(3), 1965-1982.

Kieft, M., Rijlaarsdam, G., & Van den Bergh, H. (2008). An Aptitude Treatment Interaction Approach to Writing-to-Learn. *Learning and Instruction*, 18, 379-390.

Kim, K., Hwang, J., & Zo, H. (2016). Understanding users' continuance intention toward smartphone augmented reality applications. *Information development, 32*(2), 161-174.

Kim, S.H., Park, N.H., & Joo, K.H. (2014). Effects of flipped classroom based on smart learning on self-directed and collaborative learning. International Journal of control and automation. 7(12), 69-80.

Kincheloe, J.L. (2003) *Teachers as Researchers: Qualitative Inquiry as a Path to Empowerment*. Londres: Routledge.

King, D., Greaves, F., Exeter, C., & Darzi, A. (2013). "Gamification': influencing health behaviours with games', *Journal of the Royal Society of Medicine*, 106, 76–78,

Kipper, G., & Rampolla, J. (2012). *Augmented reality*. Amstedam: Syngress.

Kirschner, P.A., & Kester, L. (2016). Towards a Research Agenda for Educational Technology Research. In VV.AA. *The Wiley Handbook of Learning Technology* (pp. 523-541). London: John Wiley & Sons

Knezek, G., Christensen, R., & Furuta, T. (2019). Validation of a teacher educator technology competencies survey. *Journal of Technology and Teacher Education*, *27*(4), 465-498.

Knowlton, J.Q. (1964). A Conceptual Scheme for Audiovisual Field. *Bulletin of the School of Education, Indiana University*. 40 (3), 1-44.

Koichiro, M. (2013). Cultivating the Ground for the Study of Education as an Inter-disciplinary Enterprise: A Philosophical Perspective. *Educational Studies in Japan: International Year Book*, 7(3), 37-49.

Krause, J.N., O'Neil, K., & Dauenhauer, B. (2017). 'Plickers: a formative assessment tool for K–12 and PETE professionals', Strategies: A Journal for Physical and Sport Educators, 30, 30–36.

Kregor, G., Padgett, L., & Brown, N. (eds.) (2013). *Technology Enhanced Learning and Teaching*. Hobart: Tasmanian Institute of Learning and Teaching, University of Tasmania.

Kuh, G.D. (2008). *High-impact educational practices: What they are, who has access to them, and why they matter*. Washington: Association of American Colleges and Universities.

Labaree, D. (2004). The Trouble with Ed Schools. New Haven and London: Yale University Press.

Laudonia, I., Mamlok-Naaman, R., Abels, S., & Eilks, I. (2018). Action research in science education–an analytical review of the literature. *Educational Action Research*, *26*(3), 480-495.

Lazarsfeld P., & Merton, R.K. (1948). Mass Communication, Popular Taste, and Organized Social Action. En B. Rosenbergand & D. M. White (Eds.). *Mass Culture: The Popular Arts in America* (pp. 457-473). Nueva York: Van Nostrand Reinhold.

Leahy, S.M., Holland, C., & Ward, F. (2019). The digital frontier: Envisioning future technologies impact on the classroom. *Futures*, 113, 102422.

Lee, A. (2010). What Count as Educational Research? Spaces, Boundaries, and Alliances. *The Australian Educational Researcher.* 37(4) 63-78.

Lehmann, J., Goussios, C., & Seufert, T. (2016). Working memory capacity and disfluency effect: an aptitude-treatment-interaction study. *Metacognition and Learning*, *11*(1), 89-105.

Lehrer, R., & Randle, L. (1987). Problem Solving, Metacognition and Composition: The Effect of Interactive Software for First-Grade Children. *Journal of Educational Computing Research*, 4, 409-427.

Lester, J.N., & Nusbaum, E.A. (2018). "Reclaiming" disability in critical qualitative research: Introduction to the special issue. *Qualitative Inquiry*, *24*(1), 3-7.

Levin, H.M. (1998). "Accelerated Schools: a decade of evolution," in *International Handbook of educational change*, Part two, eds. A. Hargreaves, A. Lieberman, M. Fullan and D. Hopkins (New York, NY: Kluwer Academic Publishers), 807-830.

Ley Orgánica 8/2013, de 9 de diciembre, para la mejora de la calidad educativa. Retrieved from https://www.boe.es/buscar/act.php?id=BOE-A-2013-12886

Liu, D., Bhagat, K.K., Gao, Y., Chang, T.W., & Huang, R. (2017). The potentials and trends of virtual reality in education. In *Virtual, augmented, and mixed realities in education* (pp. 105-130). Singapore: Springer.

Liu, D., Dede, C., Huang, R., & Richards, J. (2017). Virtual, Augmented, and Mixed Realities in Education. Singapore: Springer

Liu, G.Z., Wu, N.W., & Chen, Y.W. (2013). Identifying emerging trends for implementing learning technology in special education: A state-of-the-art review of selected articles published in 2008-2012. *Research in developmentaldisabilities*, *34*(10), 3618–3628.

Livingstone, S. (2012). Critical Reflections on the Benefits of ICT in Education. *Oxford Review of Education*, 38(1), 9-24

López Meneses, E., & Gómez Galán, J. (2010). Prácticas universitarias constructivistas e investigadoras con software social. *Praxis*, 6(1), 15-31.

López, J., & Ortega, J. M. (2017). *Revista Electrónica de Investigación y Docencia, 18*, 97-108.

López-Arenas, J.M. (1985). La Tecnología Educativa: Implicaciones para el Futuro de la Educación. *Cuestiones Pedagógicas: Revista de Ciencias de la Educación*, 2, 189-196.

López-Belmonte, J., Pozo, S., & Fuentes, A. (2019). Recursos tecno-pedagógicos de apoyo a la docencia: La realidad aumentada como herramienta dinamizadora del profesor sustituto *International Journal of Educational Research and Innovation (IJERI)*,12, 122-136

López-Belmonte, J., Pozo, S., Morales, Mª. B., & López-Meneses, E. (2019). Competencia digital de futuros docentes para efectuar un proceso de ensenanza y aprendizaje mediante realidad virtual. *EDUTEC. Revista Electrónica de Tecnología Educativa, 67*, 1-15. https://doi.org/10.21556/edutec.2019.67.1327

López-Meneses, E. (2017). El fenómeno MOOC y el futuro de la universidad. *Fronteras de la Ciencia*, 1, 90-97.

López-Meneses, E., Vázquez-Cano, E., Gómez-Galán, J., & Fernández-Márquez, E. (2019). Pedagogía de la innovación con tecnologías. Un estudio de caso en la Universidad Pablo de Olavide. *El Guiniguada. Revista de Investigaciones y Experiencias en Ciencias de la Educación*, 28, 76-92.

Lorca Marín, A.A., & Vázquez-Bernal, R.S. (2012). 'Los videojuegos para el profesorado en formación inicial de educación Infantil en la enseñanza de las Ciencias de la Naturaleza', *XXV Encuentros de Didáctica de las Ciencias Experimentales*, pp. 781-788, 2012.

Loyens, S.M., Magda, J., & Rikers, R.M. (2008). Self-directed learning in problem-based learning and its relationships with self-regulated learning. *Educational Psychology Review*, 20(4), 411-427.

Lozano, J., Ballesta, F., Cerezo, M.C., & Alcaraz, S. (2013). Las tecnologías de la información y comunicación (TIC) en el proceso de enseñanza y aprendizaje del alumnado con trastornos del espectro autista (TEA). *Revista Fuentes, 14*, 193-208.

Lu, S.J., & Liu, Y.C. (2015). Integrating augmented reality technology to enhance children's learning in marine education. *Environmental Education Research*, 21(4), 525-541.

Luque Parra, D., & Rodríguez Infante, G. (2009). Tecnologías de la Información y Comunicación aplicadas al alumno con discapacidad: un acercamiento docente. *Revista Iberoamericana de Educación, 49*(3), 8.

Luque, A. (2012). La educación inclusiva y el mundo digital: nuevos retos en la sociedad del conocimiento *Etic@*net, *2*(12), 202-215.

M. J., Berman, A.I., & Atkinson, R.C. (Ed)(2015). Past as Prologue: The

Macdonald, S.J., & Clayton, J. (2013). Back to the future, disability and the digital divide. *Disability & Society*, 28(5), 702-718.

Makino, Y. (2007). The third generation of e-learning: expansive learning mediated by a weblog. *International Journal of Web Based Communities, 3*(1), 16. Moen, A., Mørch, A. I., & Paavola, S. (Eds.). (2012). *Collaborative knowledge creation: practices, tools, concepts*. Rotterdam: Sense Publ.

Maquilón, J.J., Mirete Ruiz, A.B., & Avilés Olmos, M. (2017). La realidad aumentada (R.A). Recursos y propuestas para la innovación educativa. *Revista Electrónica Interuniversitaria de Formación del Profesorado*, 20(2), 183–203.

Markauskaite, L., & Wardak, D. (2015). Research Students' Conceptions of the Role of Information and Communication Technologies in Educational Technology Research. *Australasian Journal of Educational Technology*, 31 (4), 421-438.

Markopoulos, P., & Bekker, M. (2002). How to compare usability testing methods with children participants. *Interaction Design and Children*, 2, 28-34.

Marley, S.C., & Levin, J. R. (2011). When are Prescriptive Statements in Educational Research Justified? *Educational Psychology Review*, 23 (2), 197-206.

Martin, A., & Grudziecki J. (2006). DigEuLit: Concepts and Tools for Digital Literacy Development. *Innovation in Teaching and Learning in Information and Computer Sciences* 5, 249–267.

Martínez-Navarro, G. (2017). 'Tenologías y nuevas tendencias en educación: aprender jugando. El caso de Kahoot', 83, 252-277.

Masterman, L. (1990). *Teaching the Media*. Nueva York: Routledge.

Masterman, L. (2001) A Rationale for Media Education. En R. Kubey, (Ed). *Media Literacy in the Information Age: Current Perspectives. Information and Behaviour*. New Brunswick: Transaction Publishers.

Matheson, R. Socrative mobile quiz app saves teachers time and offers real-time data on student comprehension of material, *MIT News Office*. Retrieved from https://news.mit.edu/2014/socrative-app-real-time-data-student-comprehension-1211

Matosas-López, L., Aguado-Franco, J.C., & Gómez-Galán, J. (2019). Constructing an instrument with behavioral scales to assess teaching quality in blended learning modalities. *Journal of New Approaches in Educational Research*, 8(2), 142-165.

Mattelart, A. (1994). *World Communications: A History of Strategies and Ideas*, Minneapolis: University of Minnesota Press,

Maykut, P., & Morehouse, R. (2003). *Beginning Qualitative Research: A Philosophic and Practical Guide*. London: Falmer.

Mayor, D. (2017). Aprendizaje–Servicio: una práctica educativa que favorece la participación fuerte de los menores en la construcción de las ciudades. *Educación y Ciudad*. 33, 171–184.

Mayr, O. (1982). The Science-Technology Relationship. En B. Barnes and D. Edge (eds.), *Science in Context: Readings in the Sociology of Science* (pp. 155-63). Cambridge, MA: The MIT Press,

McKenzie, W. (2005). *Multiple Intelligences and instructional technology* (2nd ed.). Eugene, OR: International Society for Technology in Education.

McMillan, D.W., & Chavis, D.M. (1986). Sense of Community: A Definition and Theory. *J. Community Psychol*. 14, 6–23.

Mejías, A. (2008). My self-as-philosopher and My self–as-Scientist Meet to do Research in the Classroom: Some Davidsonian Notes on the Philosophy of Educational Research. *Studies in Philosophy and Education*, 27 (2-3), 161-171

Méndez, A.G., López, S., & Barra, E. (2019). Efectividad de los MOOC para docentes en el uso seguro de las TIC. *Comunicar*, 61, 103-112.

Mendía, R. (2012). El Aprendizaje-Servicio como una estrategia inclusiva para superar las barreras al aprendizaje y a la participación. Revista Educación Inclusiva. 5, 71–82.

Mendoza-Moreira, F.S., Andrade-García, B.R., Moreira-Macías, B.A., and Arteaga-Vera, J.C. (2014). Strategies for the implementation of an interactive classroom methodological approach for the formation of inverted Education degree. *Revista Educación y Tecnología.* 5, 36-48.

Mertala, P. (2020). Paradoxes of participation in the digitalization of education: a narrative account. *Learning, Media and Technology*, *45*(2), 179-192.

Mertens, D. (2005). *Research and Evaluation in Education and Psychology: Integrating Diversity with Quantitative, Qualitative, and Mixed Methods* (2nd Ed). Thousand Oaks: Sage Publications

Miles, M., & Huberman, M. (1994). *Qualitative Data Analysis: An Expanded So*

Mills, G.E. (2000). Action Research: A Guide for the Teacher Researcher. Upper Saddle River: Prentice-Hall.

Milman, N.B. (2012). The flipped classroom strategy: What is it and how can it best be used? *Distance Learning*, 9(3): 85.

Ministerio de Educación, Cultura y Deporte (2015). *Plan estratégico de aprendizaje a lo largo de la vida.* Madrid: Secretaría General Técnica.

Mirete, A.B., Cabello, F., Martínez Segura, M.J. & García Sánchez. (2013). *Cuestionario para la Evaluación de Aspectos Didácticos, Técnicos y Pedagógicos de Webs Didácticas (CEETP).* Murcia: EDITUM.

Mohajan, H. K. (2018). Qualitative research methodology in social sciences and related subjects. *Journal of Economic Development, Environment and People*, *7*(1), 23-48.

Molina & otros (2012). Las TIC en la formación inicial y en la formación permanente del profesorado de infantil y primaria. EDUTEC, *Revista Electrónica de Tecnología Educativa*, 41. Retrieved from http://edutec.rediris.es/Revelec2/Revelec41/TIC_formacion_inicial_permanente_profesorado_infantil_primaria.html

Moliner, L. (2010). El Aprendizaje Servicio en la Universidad: una estrategia en la formación de ciudadanía crítica. *Revista Electrónica Universitaria de Formación del profesorado*, 13, 69-77.

Montero, L.M., García-Salazar, J.H., & Rincón-Méndez, L.C. (2008). A Learning Experience Using Digital Enviroments: Basic Aptitudes for Everyday Living. *Educación y Educadores*, 11(1), 183-198.

Moreno-Martínez, N., & Leiva, J.J. (2017). Experiencias formativas de uso didáctico de la realidad aumentada con alumnado del grado de educación primaria en la universidad de Málaga, *Revista de Educación Mediática y TIC (Edmetic), 6* (1), 81-104.

Morlà, T. (2015). Comunidades de aprendizaje, un Sueño que hace más de 35 años que transforma realidades. *Historia Social y de la Educación*, 4, 137-162.

Muñoz, M.M., & Ayuso, M.J. (2014). Inteligencias múltiples, ¿ocho maneras diferentes de aprender? *Escuela Abierta*, 17, 103-116.

Muthanna, A. (2019). Critical qualitative inquiry and methodological awareness: The effectiveness of face-to-face interviews in changing/enhancing participants' beliefs and practices. *International Journal of Research*, *8*(2), 59-66.

Narváez Rivero, M., & Prada Mendoza, A. (2015). Self-directed learning and academic performance [Aprendizaje autodirigido y desempeño académico]. *Tiempo de Educar*. 6(11): 115-146.

Nebel, S., Schneider, S., & Rey, G. D. (2016). Mining Learning and Crafting Scientific Experiments: A Literature Review on the Use of Minecraft in Education and Research. *Educational Technology & Society,* 19 (2), 355–366.

Negroponte, N. (1995). *Being Digital.* Nueva York: Alfred A. Knopf.

Nleya, P.T.T. (2016). Transformative Applications of ICT in Education: The Case of Botswana Expansive School Transformation (Best) Project. In F. J. Mata & A. Pont (Eds.), *ICT for Promoting Human Development and Protecting the Environment* (Vol. 481, pp. 68–82). Cham: Springer International Publishing.

O'Connel, A.A., & Gray, D.L. (2011). Cause and Event: Supporting Causal Claims through Logistic Models. *Educational Psychology Review*, 23(2), 245-261

OECD. (2018). *The future of education and skills Education 2030: The future we want.* Paris: OECD.

O'Flaherty, J., & Phillips, C. (2015). The use of flipped classrooms in higher education: A scoping review. *The Internet and Higher Education.* 25, 85–95.

Orlikowski, W., & Baroudi, J.J. (1990). *Studying Information Technology in Organizations: Research Approaches and Assumptions.* Nueva York: Center for Research on Information Systems. Information Systems Department LeonardN. Stern, School of Business, New York University, Working Paper Series. STERN IS-90-4

Orozco, G. H., Tejedor, F. & Calvo, M. I. (2017). Meta-análisis sobre el efecto del software educativo en alumnos con necesidades educativas especiales. *Revista de Investigación Educativa, 35*(1), 35-52.

Ortiz, A., Almazan, L., Peñaherrera, M., & Cachón J. (2014). Formación en TIC de futuros maestros desde el análisis de la práctica en la Universidad de Jaén. *Pixel-Bit. Revista de Medios y Educación*, 44, 127-142.

Ott, B.L., & Mack, R.L. (2020). *Critical media studies: An introduction*. John Wiley & Sons.

Owen, T.R. (1999). Self-Directed Learning Readiness among Graduate Students: Implications for Orientation Programs. *Journal of College Student Development*, 40(6), 739-743.

Paavola, S., Lipponen, L., & Hakkarainen, K. (2004). Models of Innovative Knowledge Communities and Three Metaphors of Learning. *Review of Educational Research, 74*(4), 557–576.

Padilla, M. (2015). Phenomenology in educational qualitative research: Philosophy as science or philosophical science. *International Journal of Educational Excellence, 1*(2), 101-110.

Palmarini, R., Erkoyuncu, J.A., Roy, R., & Torabmostaedi, H. (2018). A systematic review of augmented reality applications in maintenance. *Robotics and Computer-Integrated Manufacturing, 49*, 215-228

Paris, S.G., & Winograd, P. (2003). The Role of Self-Regulated Learning in Contextual Teaching: Principals and Practices for Teacher Preparation. In Kenneth R. Howey (Proyect Director), *A Commissioned Paper for the U.S. Department of Education project, "Preparing Teachers to Use Contextual Teaching and Learning Strategies to Improve Student Success in and beyond School"*. Available at: http://www.ciera.org/library/archive/2001-04/0104prwn.pdf

Park, B., Münzer, S., Seufert, T., & Brünken, R. (2016). The role of spatial ability when fostering mental animation in multimedia learning: An ATI-study. *Computers in Human Behavior, 64*, 497-506.

Parkhurst, P.E. (1975). Generating Meaningful Hypotheses with Aptitude-Treatment Interactions. *AV Communication Review*, 23 (2), 171-184.

Parra, J., González, C.C., Vargas, O.L., & Saiz, J.L. (2014). Género, autodirección del aprendizaje y desempeño académico en estudiantes de pedagogía. *Educ. Educ.* 17(1), 91–107.

Paul, J. (2005). *Introduction to the Philosophies of Research and Criticism in Education and the Social Sciences*. Upper Sanddle River, New Jersey: Pearson & Merril Prentice Hall.

Pede, J. (2017). 'The effects of the online game Kahoot on science vocabulary acquisition', Theses and Dissertations, Rowan University

Pelling, N. (2015). 'Gamification Past and Present', *Conference GWC14*, Barcelona, Spain,

Peña, B., Zabalza, I., Usón, S., Llera, E. M., Martínez, A., & Romeo, L. M. (2017). "Experiencia piloto de aula invertida para mejorar el proceso de enseñanza-aprendizaje en la asignatura de Termodinámica Técnica" in In-Red 2017. *III Congreso Nacional de innovación educativa y de docencia en red* (Valencia, Spain), 583-597.

Peña, I. (2014) Utilización de MOOC en la formación docente: ventajas, desventajas y peligros Profesorado. *Revista de Currículum y Formación del Profesorado*, 18(1), 155-166.

Pennington, R.C. (2010). Computer-assisted instruction for teaching academic skills to students with autism spectrum disorders: A review of literature. *Focus on Autism and Other Developmental Disabilities*, *25*(4), 239–248.

Pérez L.M, & Ochoa A.C. (2017). El aprendizaje – servicio (APS) como estrategia para educar en ciudadanía. *Alteridad*, 12, 175-187.

Pérez Parras, J., & Gómez Galán, J. (2015). Knowledge and Influence of MOOC Courses on Initial Teacher Training. International Journal of Educational Excellence, 1(2), 81-99.

Persky, A.M., & McLaughlin, J.E. (2017). The Flipped Classroom – From Theory to Practice in Health Professional Education. American Journal of Pharmaceutical Education. 81(6), 181.

Persson, H., Åhman, H., Yngling, A.A., & Gulliksen, J. (2015). Universal design, inclusive design, accessible design, design for all: different concepts—one goal? On the concept of accessibility—historical, methodological and philosophical aspects. *Universal Access in the Information Society, 14*(4), 505-526.

Phillips, D.C. (2005). The Contested Nature of Empirical Educational Research (and Why Philosophy of Education Offers Little Help). *Journal of Philosophy of Education*, 39 (4), 577-597.

Plickers Official Website 'About us'. Available at https://www.plickers.com /about [Accessed April 11, 2018].

Plump, C. & LaRosa, J. (2017). 'Using Kahoot! in the classroom to create engagement and active learning: a game-based technology solution for eLearning novices', *Management Teaching Review*, 2, pp.151-158

Pomerol, J.C., Epelboin, Y., & Thoury, C. (2015). MOOCs: Design, Use and Business Models. New York: John Wiley & Sons.

Ponce, O.A. (2014). *Investigación de Métodos Mixtos en Educación.* Hato Rey, Puerto Rico: Publicaciones Puertorriqueñas Inc.

Ponce, O.A. (2016). *Investigación Educativa.* San Juan, Puerto Rico: Publicaciones Puertorriqueñas Inc.

Ponce, O.A., & Pagán-Maldonado, N. (2016). Investigación Educativa: Retos y Oportunidades. In J. Gómez Galán, E. López Meneses y L. Molina (Eds.). *Research Foundations of the Social Sciences* (pp. 110-121). Cupey: UMET Press.

Ponce, O.A., Gomez Galan, J., & Pagán, N. (2019). Current scientific research in the humanities and social sciences: Central issues in educational research. *European Journal of Science and Theology*, *15*(1), 81-95.

Ponce, O.A., Pagán, N.P., & Gómez Galán, J. (2017). *Filosofía de la investigación educativa en una era global: retos y oportunidades de efectividad científica.* San Juan: Publicaciones Puertorriqueña.

Ponce, O.A., Pagán, N., & Gómez-Galán, J. (2018). Research of educational policies: science over ideology. *Revista Espacios*, *39*(43).

Ponce, O.A., & Pagán-Maldonado, N. (2015). Mixed Methods Research for Education: Capturing the Complexity of the Profession. *International Journal of Educational Excellence*, 1(1), 111-135.

Popper, K.R. (1962). *La Lógica de la Investigación Científica.* Madrid: Tecnos.

Popper, K.R. (1983). Conjeturas y Refutaciones. El Desarrollo del Conocimiento Científico. Barcelona: Paidós.

Potter, W.J. (2010). The State of Media Literacy. *Journal of Broadcasting & Electronic Media*, 54 (4), 675-696.

Potter, W.J. (2013). Review of Literature on Media Literacy. *Sociology Compass*, 7 (6), 417-435.

Premarathne, P.K. (2017). 'A study on incorporating gamification into ESL classroom via Kahoot!', Proc. In *International Conference on the Humanities* (ICH), Kelaniya, Sri Lanka.

Prendes, C. (2015). Realidad aumentada y educación: análisis de experiencias prácticas, *Pixel-Bit. Revista de Medios y Educación, 46*, 187-203.

Prendes, M.P., & Gutiérrez, I. (2013). Competencias tecnológicas del profesorado en las universidades españolas. *Revista de Educación*, 361. 196-222.

Puig, J., Gijón, M., Martín, X., & Rubio, L. (2011). Aprendizaje – servicio y Educación para la Ciudadanía. *Revista de Educación*, 1, 45-67.

Puig, J.M., & Palos, J. (2006). Rasgos pedagógicos del aprendizaje – servicio. *Cuadernos de Pedagogía*, 357, 60-63.

Quizizz. Gamifying education! AngelList. Available at https://angel.co/quizizz [Accessed April 11, 2018].

Rader, H. (1990). Bibliographic instruction or information literacy. *College and Research Libraries News*, 51, 18.

Radford, M. (2006). Researching Classrooms: Complexity and Chaos. *British Educational Research Journal, 32* (2), 177–190

Radianti, J., Majchrzak, T.A., Fromm, J., & Wohlgenannt, I. (2020). A systematic review of immersive virtual reality applications for higher education: Design elements, lessons learned, and research agenda. *Computers & Education*, 147, 103778.

Ramírez, E., Cañedo, I., & Clemente, M. (2012). Las actitudes y creencias de los profesores de secundaria sobre el uso de Internet en sus clases. *Comunicar, 38*, 147-155.

Ramírez, M.S. (2013). *Competencias Docentes y Prácticas Educativas Abiertas en Educación a Distancia.* México D. F.: SINED.

Ramos, G., Chiva, I., & Gómez, Ma. B. (2017). Las competencias básicas en la nueva generación de estudiantes universitarios: Una experiencia de Innovación. *Revista de Docencia Universitaria, 15*(1), 37–55. doi:10.4995/redu.2017.5909

Rangel, P., & Peñalosa, E. (2013). Alfabetización digital en docentes de educación superior: construcción y prueba empírica de un instrumento de evaluación. *Pixel-Bit. Revista de Medios y Educación*, 43, 9-23.

Rasi, P., Vuojärvi, H., & Ruokamo, H. (2019). Media Literacy Education for All Ages. *Journal of Media Literacy Education, 11*(2), 1-19.

Rasmussen, I., & Ludvigsen, S. (2009). The Hedgehog and the Fox: A Discussion of the Approaches to the Analysis of ICT Reforms in Teacher Education of Larry Cuban and Yrjö Engeström. *Mind, Culture, and Activity, 16*(1), 83–104.

Ravitch, S.M. (2014). The Transformative Power of Taking an Inquiry Stance on Practice: Practitioner Research as Narrative and Counter-Narrative. *Perspective of Urban Education* 11(1), 1-10.

Reeves, T.C., & Reeves, P.M. (2015). Educational Technology Research in a VUCA World. *Educational Technology*, 55 (2), 26-30.

Riel, M., & Polin, L. (2004). Learning Communities: Common Ground and Critical Differences in Designing Technical Support, in *Designing for virtual communities in the service of learning*, eds. S. Barab, R. Kling and J. Gray (Cambridge, MA: Cambridge University Press), 16–50.

Rincón, E.G., Mena, J., Ramírez, M.S., & Ramírez, R. (2020). The use of gamification in xMOOCs about energy: Effects and predictive models for participants' learning. *Australasian Journal of Educational Technology*, 36(2), 43–59.

Rivas, M.R., & Rodríguez, A.B.S. (2015). Estudio sobre la intervención con Software educativo en un caso de TDAH. *Revista de Educación Inclusiva*, 8(2), 121-138.

Rivero M.P. (2017). Procesos de gamificación en el aula de ciencias sociales. *Iber: Didática de las Ciencias Sociales, Geografía e Historia*, 86, 4-6.

Rocha Seixas, L., da Gomes, A.S., & de Melo Filho, I.J. (2016). 'Effectiveness of gamification in the engagement of students', *Computers in Human Behavior*, 58, 48-63.

Rodríguez-García, A.M., Raso-Sánchez, F., & Ruiz-Palmero, J.R. (2019). Competencia digital, educación superior y formación del profesorado: un estudio de meta-análisis en la Web of Science. *Pixel-Bit. Revista de Medios y Educación,* 54, 65-81.

Rogers, E.M. (1986). *Communication Technology: The New Media in Society*. New York: Free Press.

Romero, L.M., & Troyano, J.A. (2010). *Análisis Comparativo entre las Plataforma de más Frecuente Implantación en los Sistemas Virtuales de Formación frente a un Modelo: Proyecto Sakai*. Sevilla. Universidad de Sevilla.

Romero, M., & Patiño, A. (2018). Usos pedagógicos de las TIC: del consumo a la co-creación participativa. *Revista Referencia Pedagógica*, *6*(1), 2-15.

Romero, S., González, I., & Lozano, A. (2018). Herramientas tecnológicas para la educación inclusive. *Tecnología, Ciencia y Educación*, 9, 83-111.

Roni, S.M., Merga, M.K., & Morris, J.E. (2020). *Conducting quantitative research in education*. Singapore: Springer Singapore.

Rosario, H., & Vásquez, L. (2012). Formación del docente universitario en el uso de tic. Caso de las universidades públicas y privadas. (U. de Carabobo y U. Metropolitana). *Pixel-Bit. Revista de Medios y Educación*, 41, 163-171.

Roth, W.M., & Lee, Y.J. (2006). Contradictions in theorising and implementing communities in education. *Educ. Res. Rev-Neth*. 1: 27-40.

Rowe, M., & Oltmann, C. (2016). Randomised Controlled Trials in Educational Research: Ontological and Epistemological limitations. *African Journal of Health Professions Education*, 8(1), 6-8.

Rueda, R., & Dembo, M.H. (1995). Motivational processes in learning: A comparative analysis of cognitive and sociocultural frameworks. *Advances in Motivation and Achievement*. 9, 255–289.

Salas, Q.A., Morales, M.B., Villota, W.R., & López Meneses, E. (2019). University students perceptions on the free mass training courses online. *International Journal of Educational Excellence*, 5(1), 63-77.

Saldana, J. (2013). *The Coding Manual for Qualitative Researchers*. Thousand Oaks, CA: Sage.

Salinas, J. La Investigación ante los Desafíos de los Escenarios de Aprendizaje Futuros. *Revista de Educación a Distancia*, 32, 1-12.

Sallee, M.W., & Flood, J.T. (2012). Using Qualitative Research to Bridge Research, Policy, and Practice, *Theory Into Practice*, 51. 137–144.

Salomon, G. (1981). *Communication and Education: Social and Psichological interactions*. Beverly Hills: Sage Publications.

Sánchez-Martín, J., & Dávila-Acedo, M.A. (2017). 'Just a game? Gamifying a general science class at university: Collaborative and competitive work implications', *Thinking Skills and Creativity*, 26, 51-59.

Sánchez-Vera, M.M., Solano-Fernández, I.M., & Gonzalez-Calatayud, V. (2016). FLIPPED-TIC: A Flipped Classroom experience with preservice teachers. *Revista Latinoamericana de Tecnología Educativa.*, 15(3), 69-81.

Santos, G., Ferrán, N., & Abadal, E. (2012). Recursos Educativos Abiertos: Repositorios y Uso. *El Profesional de la Información*, 21 (2), 136-145.

Savin-Baden, M., & Major, C.H. (2004). *Foundations of problem-based learning*. McGraw-Hill Education (UK).

Schoepp, K. (2005). Barriers to Technology Integration in a Technology-Rich Environment. *Learning and Teaching in Higher Education*, 2 (1), 1-24.

Schofield, J. (2007). Increasing Generalizability of Qualitative Research. En M. Hammersley (Ed). *Educational Research and Evidence-Based Practice*. Los Angeles: Sage Publications.

Schratz, M. (Ed.). (2020). *Qualitative voices in educational research*. New York: Routledge.

Schulmeister, R. (2012). *As Undercover Student in MOOCs, Keynote Campus Innovation und Jonferenztagung*. Hamburg: University of Hamburg.

SCONUL Advisory Committee on Information Literacy. (1999). Information skills in higher education: a SCONUL position paper. Prepared by the Information Skills Task Force, on behalf of SCONUL.

Scott, C. (2014). The Status of Education and its Consequences for Educational Research: An Anthropological Exploration. *Australian Journal of Education*, 54 (3), 325-340.

Selwyn, N. (2010). Looking beyond learning: notes towards the critical study of educational technology: Looking beyond learning. *Journal of Computer Assisted Learning*, *26*(1), 65–73. Veletsianos, G. (2010). Emerging Technologies in Distance Education. In G. Veletsianos (Ed.), *Emerging Technologies in Distance Education.* (pp. 3–22). Edmonton, AB: Athabasca University Press.

Serrano, T.A., Biedermann, A.M., & Santolaya, S.J. (2016). Perfil, objetivos, competencias y expectativas de futuro profesional de los estudiantes del Grado en Ingeniería en Diseño Industrial y Desarrollo de Producto de la Universidad de Zaragoza. *Revista de Docencia Universitaria*, *14*(1), 69-96. doi:10.4995/ redu.2016.5908

Seufert, T., Schütze, M., & Brünken, R. (2009). Memory Characteristics and Modality in Multimedia Learning: An Aptitude–Treatment–Interaction Study. *Learning and Instruction*, 19 (1), 28-42.

Shannon, C.E., & Weaver, W. (1949). *The Mathematical Theory of Communication*. Urbana, IL: University of Illinois.

Shapiro, N.S., & Levine, J.H. (1999). *Creating Learning Communities: A Practical Guide to Winning Support, Organizing for Change, and Implementing Programs*. Jossey-Bass Higher and Adult Education Series. Jossey-Bass Inc. Publishers, San Francisco, CA.

Shavelson, R.J. (2015). Reflections on Scientific Research in Education. In Feuer,

Sherman, W.R., & Craig, A.B. (2003). *Understanding Virtual Reality: Interface, Application and Design*. San Francisco: Kaufmann.

Siemens, G. (2005). Connectivism: A Learning Theory of the Digital Age. International *Journal of Instructional Technology and Distance Learning*, 2(1), 3-10.

Siemens, G. (2013). Massive Open Online Courses: Innovation in Education? In R. McGreal, W. Kinuthia & S. Marshall (Eds.). *Open Educational Resources: Innovation, Research and Practice* (pp. 5-15). Vancouver: Commonwealth of Learning y Athabasca University.

Siguroardóttir, A.K. (2010). Professional learning community in relation to school effectiveness. *Scand. J. Educ. Res.* 54, 395–412.

Slavin, R.E. (1996). *Every child, every school: Success for all*. Corwin Press, Thousand Oaks, CA.

Sobenis, J., & Torres, R.J. (2019). Uso de la plataforma MOODLE y su impacto en el desarrollo de competencias intelectuales. *Opuntia Brava*, 11(1), 211-216.

Sobocinski, M. (2018). *Necessary definitions for understanding gamification in education a short guide for teachers and educators.* Working paper on Researchgate. Retrieved from https://www.researchgate.net/publication/319646230

Sohn, B.K., Thomas, S.P., Greenberg, K.H., & Pollio, H.R. (2017). Hearing the voices of students and teachers: A phenomenological approach to educational research. *Qualitative Research in Education*, *6*(2), 121-148.

Solmaz, E., & Çetin, E. (2017). 'Ask-response-play-learn: students' views on gamification based interactive response systems', *Journal of Educational & Instructional Studies in the World*, 7, 28-40.

Sotiriou, S., & Bogner, F.X. (2008). Visualizing the invisible: Augmented reality as an innovative science education scheme. *Advanced Science Letters*, 1(1), 114–122.

Soto, F.J. (2007). Nuevas tecnologías y atención a la diversidad: oportunidades y retos. Retrieed from http://www.niee.ufrgs.br/eventos/CIIEE/2007/pdf/

Spector, J.M. (2016). *Foundations of Educational Technology: Integrative Approaches and Interdisciplinary Perspectives.* 2nd Ed. New York and London: Routledge.

Spector, J.M., Johnson, T.E., & Young, P.A. (2014). An Editorial on Research and Development in and with Educational Technology. Educational *Technology Research & Development*, 62 (1), 1–12

Stein, J., & Graham, C.R. (2020). *Essentials for blended learning: A standards-based guide*. London: Routledge.

Stenhouse, L. (2003). *Investigación y Desarrollo del Currículum.* Madrid: Morata. Smeyers, P. (2013). Making Sense of the Legacy of Epistemology in Education and Educational Research. *Journal of Philosophy of Education*. 47 (2), 311-331.

Sternberg, R.J., Grigorenko, E.L., Ferrari, M., & Clinkenbeard, P. (1999). A Triarchic Analysis of an Aptitude-Treatment Interaction. *European Journal of Psychological Assessment*, 15 (1), 3.

Suárez, J. & otros (2013). Las competencias del profesorado en TIC: estructura básica. *Educación XX1*. 16.1, 39-62.

Suazo-Díaz, S.N. (2006). *Inteligencias múltiples: Manual práctico para el nivel elemental.* San Juan: La Editorial, Universidad de Puerto Rico.

Sung, K. (2015). A case study on a flipped classroom in an EFL content course. *Multimedia-Assisted Language Learning*, 18(2), 159-187.

Talbert, R. (2012). Inverted classroom. *Colleagues*, 9(1), 7.

Tecnológico de Monterrey (2015). *Reporte EduTrends. Radar de Innovación Educativa 2015.* Monterrey: Tecnológico de Monterrey.

Terigi, F. (2013). *VIII Foro Latinoamericano de Educación: saberes docentes: qué debe saber un docente y por qué*. Buenos Aires: Santillana.

Testaceni, G. (2016). MOOC: Un nuevo modelo de aprendizaje colaborativo, abierto y conectado. *Tecnología Educativa*, 1(1), 1-6.

Texas Computer Science: Home. (2019). Retrieved from http://texascomputerscience.weebly.com/flipped-classroom.html

Thomas, G., & Pring, R. (2004) *Evidence-Based Practice in Education*. Maidenhead: Open University Press.

Thompson, C. (2012). Theorizing Education and Educational Research. *Studies in Philosophy and Education*, 31(3), 239-250.

Tinto, V. (2003). "Learning Better Together: The Impact of Learning Communities on Student Success," in *Promoting Student Success in College*, Higher Education Monograph Series, (Syracuse, NY: Syracuse University), 1-8.

Tissenbaum, M., & Slotta, J.D. (2019). Developing a smart classroom infrastructure to support real-time student collaboration and inquiry: a 4-year design study. *Instructional Science*, 47 (4), 423-462 doi: https://doi.org/10.1007/s11251-019-09486-1

Toledo, P., Sánchez, J.M., & Gutiérrez, J.J. (2013). Evolución de la accesibilidad web en las Universidades Andaluzas. *Pixel-Bit. Revista de Medios y Educación*, *43*, 65-83.

Torre, D., & Murphy, J. (2015) A different lens: Changing perspectives using Photo-Elicitation Interviews. *Education Policy Analysis Archives, 23*(111).

Trimmer, K. (Ed.). (2016). *Political pressures on educational and social research: International perspectives*. London: Routledge.

Tsay, C.H.-H Kofinas, A. & Luo, J. (2018). 'Enhancing student learning experience with technology-mediated gamification: An empirical study', *Computers and Education*, 121, 1-17.

Ulvik, M., Riese, H., & Roness, D. (2018). Action research–connecting practice and theory. *Educational action research*, *26*(2), 273-287.

UNESCO (2014). UNESCO Education Strategy 2014–2021. Paris: UNESCO.

UNESCO Institute for Statistics, Law, N., Woo, D., Torre J de la, Wong G. (2018). *A Global Framework of Reference on Digital Literacy Skills for Indicator 4.4.2: UIS/2018/ICT/IP/51*. Montreal: UIS.

UNESCO (2012). Information for All-Programme (IFAP). *Available at https://en.unesco.org/programme/ifap* (accessed December 1, 2018).

Valdivieso, T.S. & González, M.A. (2016). Competencia digital docente: ¿Dónde estamos? Perfil del docente de educación primaria y secundaria. El caso de Ecuador. *Revista de Medios y Educación. 49*, 57-73.

Van Dijk, J. (2020). *The network society*. London: Sage Publications Limited.

Van Driel, J.H., Beijaard, D., Verloop, N. (2001). Professional development and reform in science education: The role of teachers' practical knowledge. *J. Res. Sci. Teach.* 38, 137–158.

Vázquez Cano, E., López Meneses, E., & Sarasola, J.L. (2013). *La Expansión del Conocimiento en Abierto: MOOCs*. Barcelona: Octaedro.

Vázquez-Cano, E., Reyes, M., Colmenares, L., & López-Meneses, E. (2017). Competencia digital del alumnado de la Universidad Católica de Santiago de Guayaquil. *Revista opción, 83*, 229-251.

Vázquez-Cano, E., López Meneses, E., & Sánchez-Serrano, J.L. (2015). Analysis of social worker and educator's areas of intervention through multimedia concept maps and online discussion forums higher Education. *Electronic Journal of e-Learning, 13*(5), 333-346

Veletsianos, G. (Ed.). (2016). *Emergence and Innovation in Digital Learning: Foundations and Applications*. Edmonton, AB: Athabasca University Press.

Vergara, D., Gómez, A.I., & Fernández, P. (2017). 'Evolución histórica de la gamificación educativa', Proc. In *IV CIFCYT Congreso Iberoamericano de Filosofía de la Ciencia y de la Tecnología*, Salamanca, Spain,

Vesisenaho, M., Juntunen, M., Häkkinen, P., Pöysä-Tarhonen, J., Fagerlund, J., Miakush, I., & Parviainen, T. (2019). Virtual reality in education: Focus on the role of emotions and physiological reactivity. *Journal of Virtual Worlds Research*, 12(1). doi:10.4101/jvwr.v12i1.7329

Veytia M. G., Gómez-Galán, J., & Cevallos, M.B. (2019). Competencias investigativas y mediación tecnológica en doctorandos de Iberoamérica. *IJERI: International Journal of Educational Research and Innovation*, 12, 1-19.

Vila, G., Castro, S., Barreiro, B., & Losada, F. (2016). Aprendizaje – Servicio en la gestión empresarial. *Revista Internacional de Investigación e Innovación en Didáctica de las Humanidades y las Ciencias*, 3, 139-149.

Vladimirovna, S. & Sergeevna, O. (2015). Features of the Information and Communication Technology Application by the Subjects of Special Education. *International Education Studies, 8*(6), doi:10.5539/ies.v8n6p162.

Vrasidas, C. (2015). The Rhetoric of Reform and Teachers' Use of ICT. *British Journal of Educational Technology*, 46 (2), 370-380.

Vygotski, L.S. (1984). Learning and intellectual development at school age [Aprendizaje y desarrollo intelectual en la edad escolar]. *Infancia y aprendizaje*, 7(27-28), 105-116.

Wallace, T., & Georgina, D. (2014). Preparing special education teachers to use educational tecnology to enhance student learning. *11th International Conference on Cognition and Exploratory Learning in Digital Age*.

Wallerstein, I.M. (2001). *Unthinking Social Science: The Limits of Nineteenth-Century Paradigms*. Filadelfia: Temple University Press.

Wang, A.I. (2015). 'The wear out effect of a game-based student response system', *Computers & Education*, 82, 217-227.

Weiss, M.J., Visher, M.G., Weissman, E., & Wathington, H. (2015). The impact of learning communities for students in developmental education: A synthesis of findings from randomized trials at six community colleges. *Educ. Eval. Policy An.*, 37, 520-541.

Wellington, J. (2015). *Educational Research: Contemporary Issues and Practical Approaches*. London: Bloomsbury Publishing.

Werbach, K., & Hunter, D. (2012). *'For the win: how game thinking can revolutionize your business'*, Pennsylvania, Wharton Digital Press

White, I., Smigiel, H., & Levin, J. (2017). Modelling the complexity of technology adoption in higher education teaching practice. In *Higher Education Technology Agenda 2017 Biennial Conference* (pp. 7-10). Auckland: Author.

Williams, C., & Beam, S. (2019). Technology and writing: Review of research. *Computers & Education*, *128*, 227-242.

Wilson, B.G., Ludwig-Hardman, S., Thornam, C.L., Dunlap, J.C. (2004). Bounded community: Designing and facilitating learning communities in formal courses. *The Int. Rev. Res. Open Dis.*, 5, 1–22.

Wilson, B., & Ryder, M. (1996). *Dynamic Learning Communities: An Alternative to Designed Instructional Systems*. Presented at Selected Research and Development Presentations at the 1996 National Convention of the Association for Educational Communications and Technology 1996. Available at: http://www.learntechlib.org/p/80071

Wood, A. F., & Smith, M.J. (2005). *Online Communication. Linking Technology, Identity & Culture*. New York and London: Lawrence Erlbaum Associates.

Wright, R.E., & Reeves, J.L. (2019). Open educational resource (OER) adoption in higher education: Examining institutional perspectives. *FDLA Journal*, 4(1), 9.

Wu, H.K., Lee, S.W.Y., Chang, H.Y., & Liang, J.C. (2013). Current status, opportunities and challenges of augmented reality in education. *Computers & Education, 62*, 41-49.

Xu, F., Weber, J., & Buhalis, D. (2014). 'Gamification in tourism'. In: Z. Xiang, I. Tussyadiah (eds), *Information and Communication Technologies in Tourism*, Cham: Springer.

Yang, S.J.H. (2006). Context aware ubiquitous learning environments for peer-to-peer collaborative learning. *Educational Technology & Society, 9*(1), 188-201.

Yarnit, M. (2000). *Towns, cities and regions in the learning age. A survey of learning communities*. London: LGA Publications for the DfEE. NCA.

Yarnit, M. (2001). "Learning communities: the secret to their success". In *Learning Communities: strengthening lifelong through practice*. Seminar 2: Local organisations and community learning. Back Ground Paper.

Yeh, Y.C., & Lin, C.F. (2015). Aptitude-Treatment Interactions during Creativity Training in E-Learning: How Meaning-Making, Self-Regulation, and Knowledge Management Influence Creativity. *Educational Technology & Society*, 18 (1), 119-131.

Yildirim, I. (2017). 'The effects of gamification-based teaching practices on student achievement and students' attitudes toward lessons', *The Internet and Higher Education*, 33, 86-92.

Ying, L., & Li, L. (2010). Augmented Reality for remote education. *Advanced Computer Theory and Engineering*, 3(3), 187-191.

Yong, J., & Yoon, H. (2015). Social Facets of Knowledge Creation: The Validation of Knowledge Assets. *Social Behavior and Personality*. 43 (5), 815-828.

Yuan, L., & Powell, S. (2013). *MOOCs and Open Education: Implications for Higher Education*. Bolton: CETIS-University of Bolton.

Yusoff, Z., Kamsin, A., Shamshirband, S., Chronopoulos, A.T. (2018). 'A survey of educational games as interaction design tools for affective learning: thematic analysis taxonomy', *Education and Information Technologies*, 23, 393-418.

Zabolotniaia, M., Cheng, Z., Dorozhkin, E., & Lyzhin, A. (2020). Use of the LMS Moodle for an Effective Implementation of an Innovative Policy in Higher Educational Institutions. *International Journal of Emerging Technologies in Learning*, 15(13), 172-189.

Zapata, M. (2013). MOOCs, una visión crítica y una alternativa complementaria: La individualización del aprendizaje y de la ayuda pedagógica. *Campus Virtuales*, 2(1), 20-38.

Zhang, L. (2015). Teaching model design of business English based on flipped classroom case study. In *Proceedings of the International Conference on Education* (pp. 176-181). Hong Kong, Management and Computing Technology.

Zichermann, G., & Linder, J. (2010). *'Game-based marketing: inspire customers loyalty through rewards, challenges and contests'*, Hoboken, NJ: John Wiley & Sons.

Zurkowski, P.G. (1974). *The information service environment relationships and priorities*. Washington: National Commission on Libraries.

Index

About the Editor

José Gómez Galán is currently Professor of Theory and History of Education at the University of Extremadura, Spain, and Research Professor at the Ana G. Mendez University, Cupey campus, Puerto Rico-USA. Among its different lines of research, those related to educational innovation currently stand out. Researcher and Visiting Professor at several international universities: University of Oxford (UK), University of Minnesota (USA), La Sapienza of Rome (Italy), several Latin American universities, etc. He is Doctor (PhD) in Philosophy and Education, Doctor (PhD) in Geography and History and holds Masters/DEA degrees and other academic degrees in History, Education, Religion, Biology, etc. in various national and international universities. He is a director and member of various research groups in different academic centers on an international scale. Director in several research projects in prestigious centers and universities in Europe (UK, France, Italy, etc.) and USA. National Educational Research Award and Academic Excellence Award (Spain). He is currently a member of committees (editorial, scientific and advisory) and/or referee of important international journals included in the JCR. His main research focus on the scientific fields of Education, Humanities and Social Sciences, in news lines of interdisciplinary dialogue.

For Product Safety Concerns and Information please contact our EU representative GPSR@taylorandfrancis.com
Taylor & Francis Verlag GmbH, Kaufingerstraße 24, 80331 München, Germany

www.ingramcontent.com/pod-product-compliance
Ingram Content Group UK Ltd.
Pitfield, Milton Keynes, MK11 3LW, UK
UKHW022210190926
475702UK00003B/288

9788770221986